STUDENT LECTURE NOTEBOOK

GENERAL CHEMISTRY

FOURTH EDITION

HILL • PETRUCCI • McCREARY • PERRY

PEARSON
Prentice Hall

Upper Saddle River, NJ 07458

Project Manager: Kristen Kaiser
Senior Editor: Kent Porter-Hamann
Editor-in-Chief, Science: John Challice
Vice President of Production & Manufacturing: David W. Riccardi
Executive Managing Editor: Kathleen Schiaparelli
Assistant Managing Editor: Becca Richter
Production Editor: Elizabeth Klug
Supplement Cover Manager: Paul Gourhan
Supplement Cover Designer: Joanne Alexandris
Manufacturing Buyer: Ilene Kahn
Cover Image Credit: MgO(100) surface /Royce Copenheaver

© 2005 Pearson Education, Inc.
Pearson Prentice Hall
Pearson Education, Inc.
Upper Saddle River, NJ 07458

Pearson Prentice Hall® is a trademark of Pearson Education, Inc.

The author and publisher of this book have used their best efforts in preparing this book. These efforts include the development, research, and testing of the theories and programs to determine their effectiveness. The author and publisher make no warranty of any kind, expressed or implied, with regard to these programs or the documentation contained in this book. The author and publisher shall not be liable in any event for incidental or consequential damages in connection with, or arising out of, the furnishing, performance, or use of these programs.

Printed in the United States of America

10 9 8 7 6 5 4 3 2 1

ISBN 0-13-146996-7

Pearson Education Ltd., *London*
Pearson Education Australia Pty. Ltd., *Sydney*
Pearson Education Singapore, Pte. Ltd.
Pearson Education North Asia Ltd., *Hong Kong*
Pearson Education Canada, Inc., *Toronto*
Pearson Educación de Mexico, S.A. de C.V.
Pearson Education—Japan, *Tokyo*
Pearson Education Malaysia, Pte. Ltd.

TABLE OF CONTENTS

Chapter 11

Chapter 12

Chapter 13

Chapter 14

Chapter 1

Chemistry: Matter and Measurement

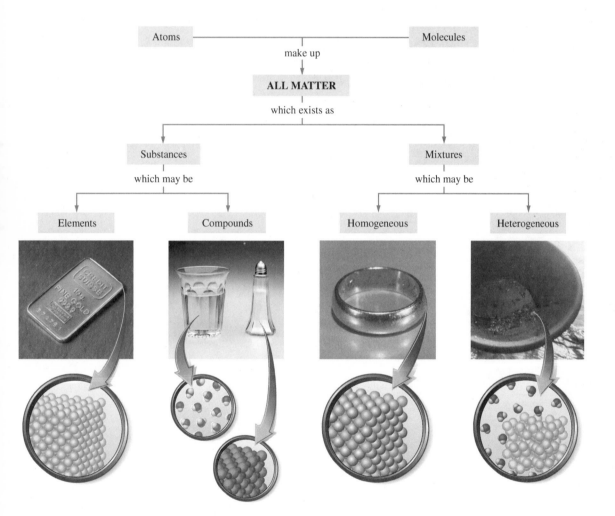

FIGURE 1.3 A scheme for classifying matter

Table 1.4
Some Common SI Prefixes

Multiple	Prefix
10^{12}	*tera* (T)
10^{9}	*giga* (G)
10^{6}	*mega* (M)
10^{3}	*kilo* (k)
10^{2}	*hecto* (h)
10^{1}	*deca* (da)
10^{-1}	*deci* (d)
10^{-2}	*centi* (c)
10^{-3}	*milli* (m)
10^{-6}	*micro* (μ)*
10^{-9}	*nano* (n)
10^{-12}	*pico* (p)

* The Greek letter μ (spelled "mu" and pronounced "mew").

TABLE 1.4 Some Common SI Prefixes

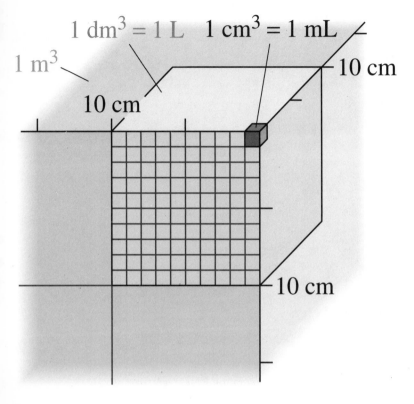

FIGURE 1.5   Some volume units compared

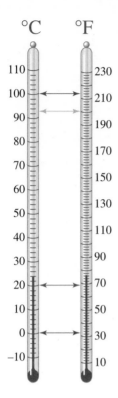

FIGURE 1.7   The Celsius and Fahrenheit temperature scales compared

(a) Low accuracy
Low precision

(b) Low accuracy
High precision

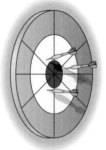

(c) High accuracy
Low precision

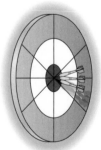

(d) High accuracy
High precision

FIGURE 1.9   Comparing precision and accuracy: a dartboard analogy

Table 1.6 Some Conversions Between Common (U.S.) and Metric Units

Metric	Common

Mass

1 kg   = 2.205 lb
453.6 g = 1 lb
28.35 g = 1 ounce (oz)

Length

1 m   = 39.37 in.
1 km   = 0.6214 mi
2.54 cm = 1 in.*

Volume

1 L   = 1.057 qt
3.785 L = 1 gal
29.57 mL = 1 fluid ounce (fl oz)

* U.S. inch is defined as exactly 2.54 cm. The other equivalencies are rounded off.

TABLE 1.6 Some conversions between common US and metric units

Chapter 2

Atoms, Molecules, and Ions

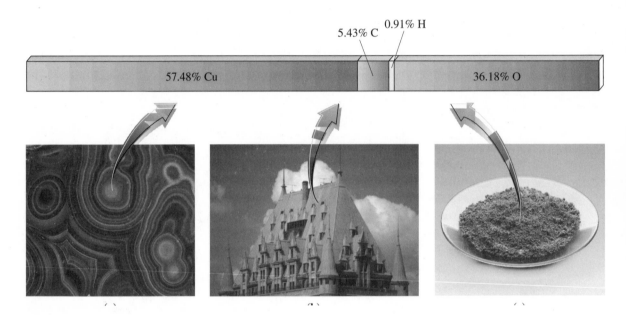

FIGURE 2.1   The law of definite proportions

	Carbon monoxide (CO)	**Carbon dioxide (CO$_2$)**
The elements	3.0 g carbon (C) + 4.0 g oxygen (O)	3.0 g carbon (C) + 8.0 g oxygen (O)
The compound	7.0 g carbon monoxide (CO)	11.0 g carbon dioxide (CO$_2$)
Oxygen-to-carbon mass ratio	$\dfrac{4.0 \text{ g oxygen}}{3.0 \text{ g carbon}}$	$\dfrac{8.0 \text{ g oxygen}}{3.0 \text{ g carbon}}$

Comparing two mass ratios:

Mass ratio for CO$_2$ $\longrightarrow$

Mass ratio for CO $\longrightarrow$

$$\frac{\dfrac{8.0 \text{ g oxygen}}{3.0 \text{ g carbon}}}{\dfrac{4.0 \text{ g oxygen}}{3.0 \text{ g carbon}}} = \frac{8.0 \text{ g oxygen}}{4.0 \text{ g oxygen}} = 2:1$$

FIGURE 2.2   The law of multiple proportions

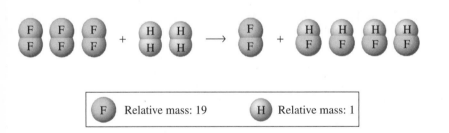

| F | Relative mass: 19 | H | Relative mass: 1 |

FIGURE 2.3   Dalton's atomic theory and the laws of constant composition and conservation of mass

Metals
Nonmetals
Noble gases

26 — Atomic number, Z
Fe — Chemical symbol
55.847 — Atomic mass (weighted average)

1A																		8A
1	H	2A											3A	4A	5A	6A	7A	He
2	Li	Be											B	C	N	O	F	Ne
3	Na	Mg	3B	4B	5B	6B	7B	—8B—			1B	2B	Al	Si	P	S	Cl	Ar
4	K	Ca	Sc	Ti	V	Cr	Mn	Fe	Co	Ni	Cu	Zn	Ga	Ge	As	Se	Br	Kr
5	Rb	Sr	Y	Zr	Nb	Mo	Tc	Ru	Rh	Pd	Ag	Cd	In	Sn	Sb	Te	I	Xe
6	Cs	Ba	La*	Hf	Ta	W	Re	Os	Ir	Pt	Au	Hg	Tl	Pb	Bi	Po	At	Rn
7	Fr	Ra	Ac†	Rf	Db	Sg	Bh	Hs	Mt	Ds	**	**						

Period

*Lanthanide series	Ce	Pr	Nd	Pm	Sm	Eu	Gd	Tb	Dy	Ho	Er	Tm	Yb	Lu
†Actinide series	Th	Pa	U	Np	Pu	Am	Cm	Bk	Cf	Es	Fm	Md	No	Lr

** Not yet named

FIGURE 2.5 The modern periodic table

1A	2A	3B	4B	5B	6B	7B	8B			1B	2B	3A	4A	5A	6A	7A	8A
Li^+														N^{3-}	O^{2-}	F^-	
Na^+	Mg^{2+}											Al^{3+}		P^{3-}	S^{2-}	Cl^-	
K^+	Ca^{2+}	Sc^{3+}	Ti^{2+} Ti^{4+}	V^{2+} V^{3+}	Cr^{2+} Cr^{3+}	Mn^{2+} Mn^{4+}	Fe^{2+} Fe^{3+}	Co^{2+} Co^{3+}	Ni^{2+}	Cu^+ Cu^{2+}	Zn^{2+}				Se^{2-}	Br^-	
Rb^+	Sr^{2+}									Ag^+	Cd^{2+}		Sn^{2+}			I^-	
Cs^+	Ba^{2+}									Au^+ Au^{3+}			Pb^{2+}				

FIGURE 2.10 Symbols and periodic table locations of some monatomic ions

Table 2.4 Some Common Polyatomic Ions

Name	Formula	Typical Compound
Cation		
Ammonium ion	NH_4^+	NH_4Cl
Anions		
Acetate ion	[a]$C_2H_3O_2^-$	$NaC_2H_3O_2$
Carbonate ion	CO_3^{2-}	Li_2CO_3
Hydrogen carbonate ion (or bicarbonate ion)[b]	HCO_3^-	$NaHCO_3$
Hypochlorite ion	ClO^-	$Ca(ClO)_2$
Chlorite ion	ClO_2^-	$NaClO_2$
Chlorate ion	ClO_3^-	$NaClO_3$
Perchlorate ion	ClO_4^-	$KClO_4$
Chromate ion	CrO_4^{2-}	K_2CrO_4
Dichromate ion	$Cr_2O_7^{2-}$	$(NH_4)_2Cr_2O_7$
Cyanate ion	OCN^-	$KOCN$
Thiocyanate ion[c]	SCN^-	$KSCN$
Cyanide ion	CN^-	KCN
Hydroxide ion	OH^-	$NaOH$
Nitrite ion	NO_2^-	$NaNO_2$
Nitrate ion	NO_3^-	$NaNO_3$
Oxalate ion	$C_2O_4^{2-}$	CaC_2O_4
Permanganate ion	MnO_4^-	$KMnO_4$
Phosphate ion	PO_4^{3-}	Na_3PO_4
Hydrogen phosphate ion	HPO_4^{2-}	Na_2HPO_4
Dihydrogen phosphate ion	$H_2PO_4^-$	NaH_2PO_4
Sulfite ion	SO_3^{2-}	Na_2SO_3
Hydrogen sulfite ion (or bisulfite ion)[b]	HSO_3^-	$NaHSO_3$
Sulfate ion	SO_4^{2-}	Na_2SO_4
Hydrogen sulfate ion (or bisulfate ion)[b]	HSO_4^-	$NaHSO_4$
Thiosulfate ion[c]	$S_2O_3^{2-}$	$Na_2S_2O_3$

[a] The acetate ion is also represented as CH_3COO^-.   [b] The prefix "bi-" means that the ion contains a replaceable H atom. This should not be confused with the prefix "di-," which means two (usually used to represent a doubling of a simpler unit).   [c] The prefix "thio-" means that a sulfur atom has replaced an oxygen atom.

TABLE 2.4   Some Common Polyatomic Ions

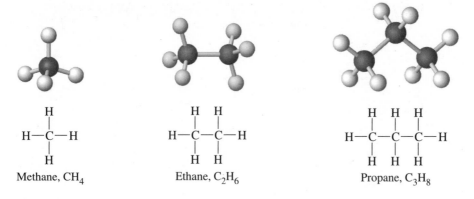

Methane, CH₄

Ethane, C₂H₆

Propane, C₃H₈

FIGURE 2.13 Alkanes

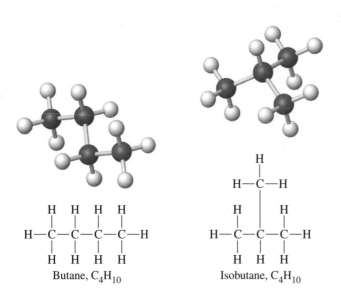

Butane, C₄H₁₀

Isobutane, C₄H₁₀

FIGURE 2.14 Butane and isobutane

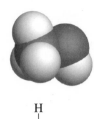

H
|
H—C—O—H
|
H

CH₃OH
Methyl alcohol (methanol)

H   H
| |
H—C—C—O—H
| |
H   H

CH₃CH₂OH
Ethyl alcohol (ethanol)

FIGURE 2.16   Alcohols

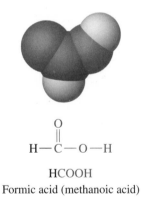

O
‖
H—C—O—H

HCOOH
Formic acid (methanoic acid)

H   O
| ‖
H—C—C—O—H
|
H

CH₃COOH
Acetic acid (ethanoic acid)

FIGURE 2.19   Carboxylic acids

Chapter 3

Stoichiometry: Chemical Calculations

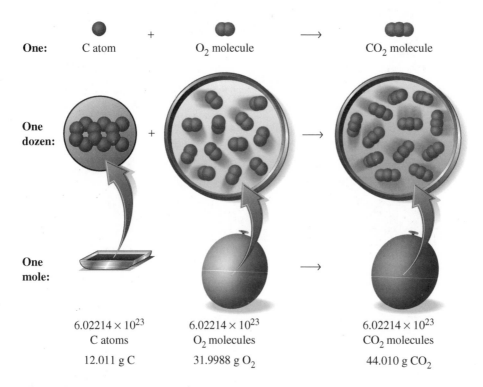

One: C atom + O_2 molecule $\longrightarrow$ CO_2 molecule

One dozen:

One mole:

6.02214×10^{23}
C atoms

6.02214×10^{23}
O_2 molecules

6.02214×10^{23}
CO_2 molecules

12.011 g C

31.9988 g O_2

44.010 g CO_2

FIGURE 3.1 Microscopic and macroscopic views of the combination of carbon and oxygen to form carbon dioxide.

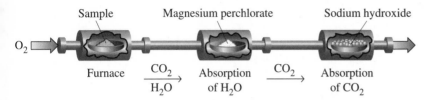

FIGURE 3.4   Apparatus for combustion analysis

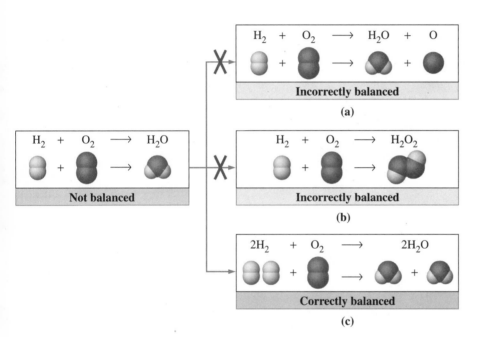

FIGURE 3.6   Balancing the chemical equation for the reaction of hydrogen with oxygen to form water

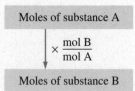

FIGURE 3.8   The mole ratio in stoichiometry

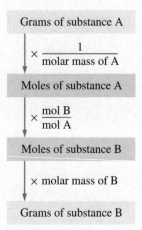

FIGURE 3.9   Stoichiometry involving mass

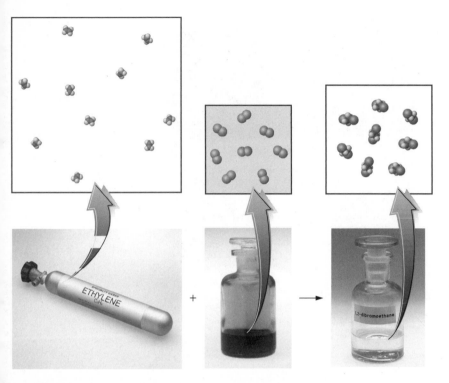

FIGURE 3.10   A molecular view of the reactants in the reaction between ethylene and bromine

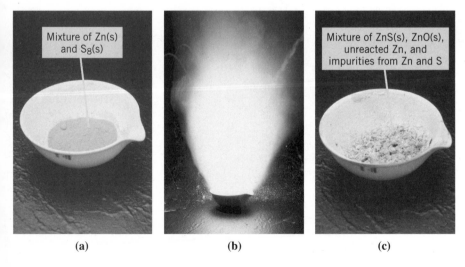

FIGURE 3.11   A reaction that has less than 100% yield

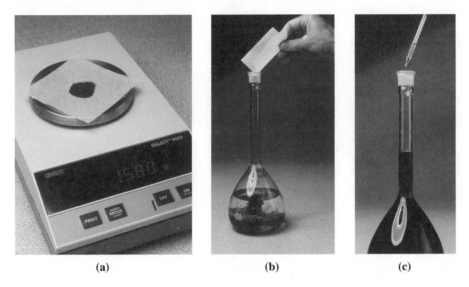

(a)   (b)   (c)

FIGURE 3.12 Preparation of 0.01000 M $KMnO_4$ solution

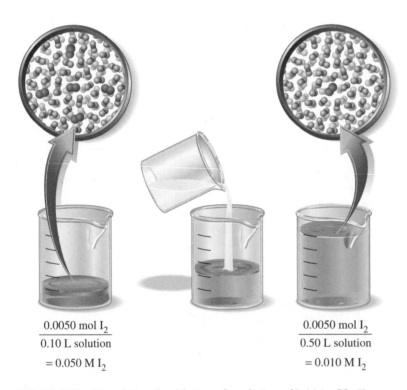

$$\frac{0.0050 \text{ mol } I_2}{0.10 \text{ L solution}}$$

$= 0.050 \text{ M } I_2$

$$\frac{0.0050 \text{ mol } I_2}{0.50 \text{ L solution}}$$

$= 0.010 \text{ M } I_2$

FIGURE 3.13 Visualizing the dilution of a solution of I_2 (s) in CS_2 (i)

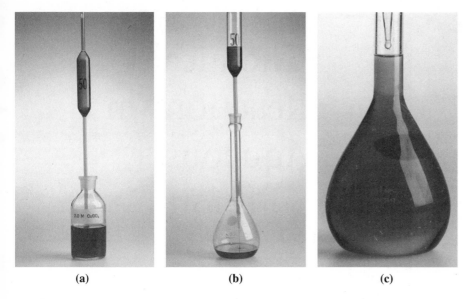

(a)   **(b)**   **(c)**

FIGURE 3.14   Dilution of a copper(II) sulfate solution: Example 3.26 illustrated

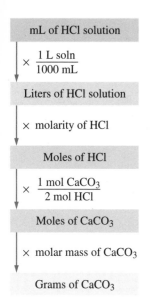

FIGURE 3.15   Stoichiometry and solutions: Flow chart for Example 3.27

Chapter 4

Chemical Reactions in Aqueous Solutions

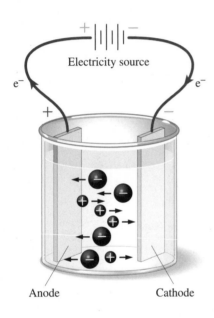

Anode

Cathode

FIGURE 4.2   Conduction of electric current through a solution

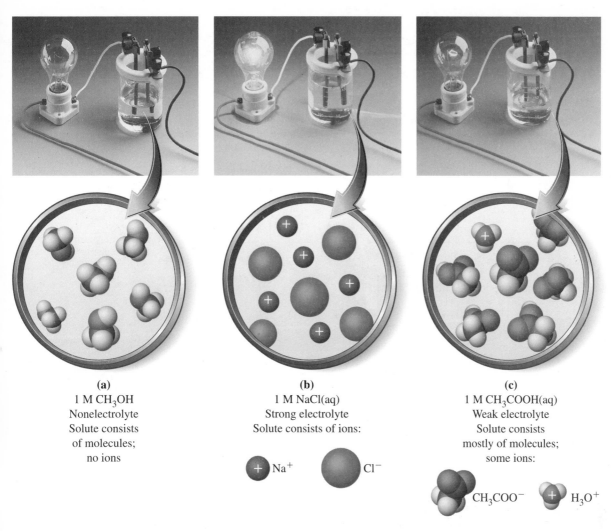

(a)
1 M CH$_3$OH
Nonelectrolyte
Solute consists
of molecules;
no ions

(b)
1 M NaCl(aq)
Strong electrolyte
Solute consists of ions:

+ Na$^+$ Cl$^-$

(c)
1 M CH$_3$COOH(aq)
Weak electrolyte
Solute consists
mostly of molecules;
some ions:

CH$_3$COO$^-$ + H$_3$O$^+$

FIGURE 4.3   Electrolytic properties of aqueous solutions

Table 4.1 Common Strong Acids and Strong Bases

Acids		Bases	
Binary Hydrogen Compounds	Oxoacids	Group 1A hydroxides	Group 2A hydroxides
HCl	HNO_3	LiOH	$Mg(OH)_2$
HBr	H_2SO_4[a]	NaOH	$Ca(OH)_2$
HI	$HClO_4$	KOH	$Sr(OH)_2$
		RbOH	$Ba(OH)_2$
		CsOH	

[a] H_2SO_4 is a strong acid in its first ionization step but weak in its second ionization step.

TABLE 4.1 Common Strong Acids and Strong Bases

Table 4.2 Some Common Gas-Forming Acid–Base Reactions

Anion	Reaction with H^+
HCO_3^-	$HCO_3^- + H^+ \longrightarrow CO_2(g) + H_2O(l)$
CO_3^{2-}	$CO_3^{2-} + 2\,H^+ \longrightarrow CO_2(g) + H_2O(l)$
HSO_3^-	$HSO_3^- + H^+ \longrightarrow SO_2(g) + H_2O(l)$
SO_3^{2-}	$SO_3^{2-} + 2\,H^+ \longrightarrow SO_2(g) + H_2O(l)$
HS^-	$HS^- + H^+ \longrightarrow H_2S(g)$
S^{2-}	$S^{2-} + 2\,H^+ \longrightarrow H_2S(g)$

TABLE 4.2 Some Common Gas-Forming Acid-Base Reactions

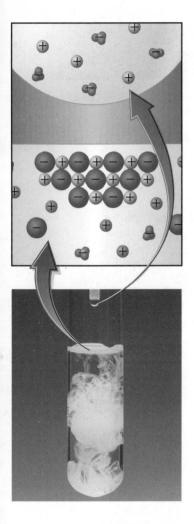

FIGURE 4.7 The precipitation of silver iodide, AgI(s)

Table 4.3   General Guidelines for the Water Solubilities of Common Ionic Compounds

Almost all nitrates, acetates, perchlorates, group 1A metal salts, and ammonium salts are *SOLUBLE*.

Most chlorides, bromides, and iodides are *SOLUBLE*. Exceptions: those of Pb^{2+}, Ag^+, and Hg_2^{2+}.

Most sulfates are *SOLUBLE*. Exceptions: those of Sr^{2+}, Ba^{2+}, Pb^{2+}, and Hg_2^{2+} ($CaSO_4$ is slightly soluble).

Most carbonates, hydroxides, phosphates, and sulfides are *INSOLUBLE*. Exceptions: ammonium and group 1A metal salts of any of those anions are soluble; hydroxides and sulfides of Ca^{2+}, Sr^{2+}, and Ba^{2+} are slightly to moderately soluble.

TABLE 4.3   General Guidelines for the Water Solubilities of Common Ionic Compounds

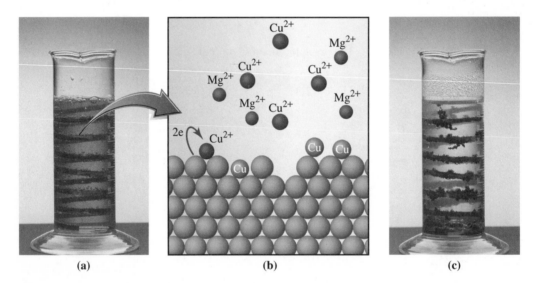

(a)   (b)   (c)

FIGURE 4.11   An oxidation-reduction reaction

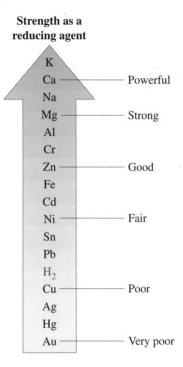

Strength as a reducing agent

K
Ca —— Powerful
Na
Mg —— Strong
Al
Cr
Zn —— Good
Fe
Cd
Ni —— Fair
Sn
Pb
H_2
Cu —— Poor
Ag
Hg
Au —— Very poor

FIGURE 4.13   Activity series of some metals

(a)   (b)   (c)

FIGURE 4.16   The Technique of Titration

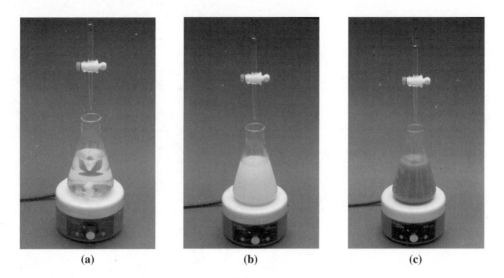

FIGURE 4.17 A precipitation titration

FIGURE 4.18 A redox titration using permanganate ion as an oxidizing agent

Chapter 5

Gases

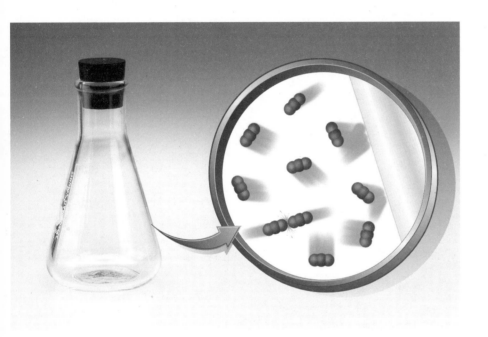

FIGURE 5.1 Visualizing molecular motion in a gas

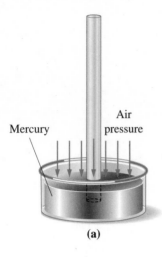

(a)

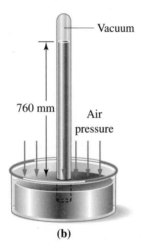

Vacuum

760 mm

Air
pressure

(b)

FIGURE 5.2   Measuring air pressure with a mercury barometer

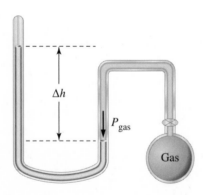

Δh

P_{gas}

Gas

FIGURE 5.3   Measuring gas pressure with a closed-end manometer

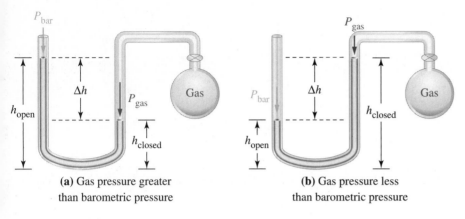

(a) Gas pressure greater
than barometric pressure

(b) Gas pressure less
than barometric pressure

FIGURE 5.4   Measuring gas pressure with an open-end manometer

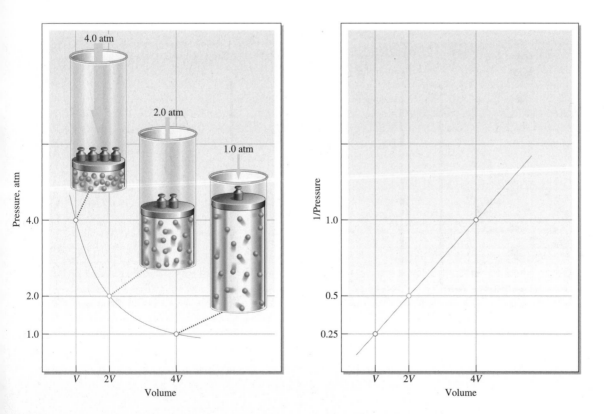

FIGURE 5.6   Boyle's law: A kinetic-theory view and a graphical representation

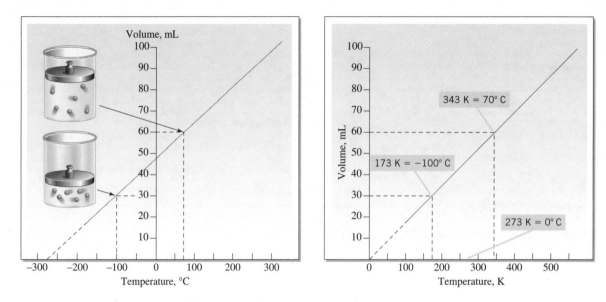

FIGURE 5.8   Charles's law: Gas volume as a function of temperature

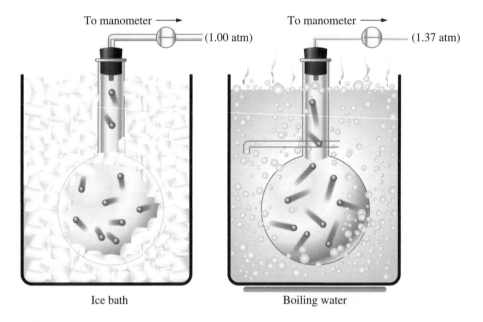

FIGURE 5.11   The effect of temperature on the pressure of a fixed amount of gas in a constant volume: A kinetic theory interpretation

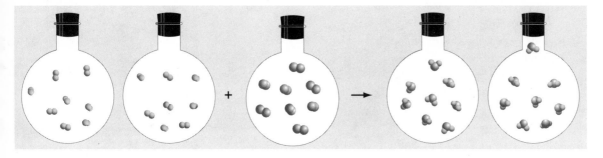

FIGURE 5.12 Avogadro's explanation of Gay-Lussac's law of combining volumes

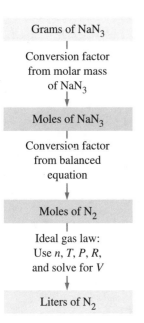

FIGURE 5.13 Stoichiometry diagram for Example 5.17

(a) 5.0 L at 20 °C   **(b)** 5.0 L at 20 °C   **(c)** 5.0 L at 20 °C

FIGURE 5.14   Dalton's law of partial pressures illustrated

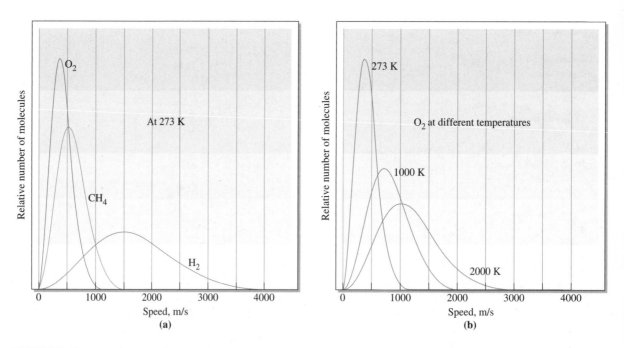

FIGURE 5.17   Distribution of molecular speeds: Effects of molar mass and temperature

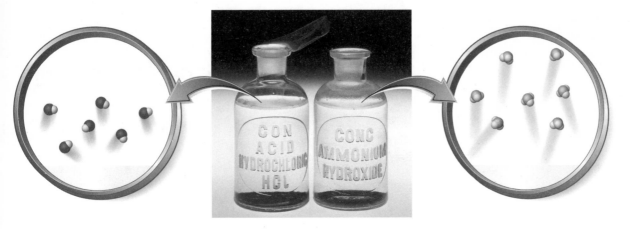

FIGURE 5.18   Diffusion of gases

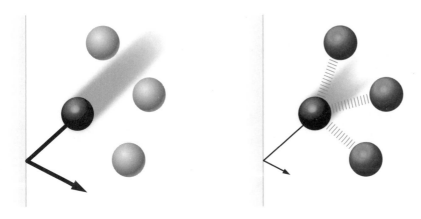

FIGURE 5.19   Pressure and intermolecular forces of attractions

Chapter 6

Thermochemistry

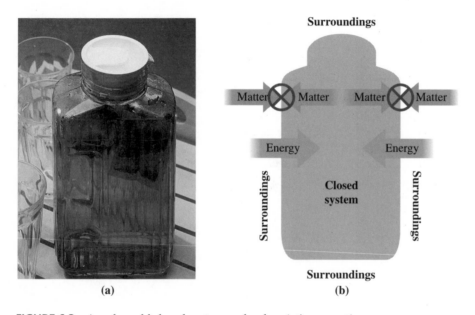

FIGURE 6.3 A real-world closed system and a chemist's conception

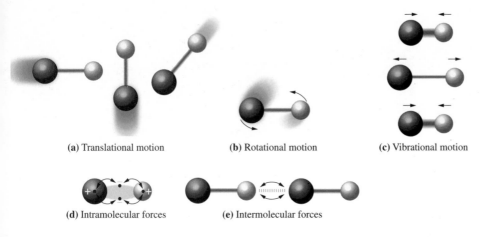

(a) Translational motion   (b) Rotational motion   (c) Vibrational motion

(d) Intramolecular forces   (e) Intermolecular forces

FIGURE 6.4   Some components of internal energy

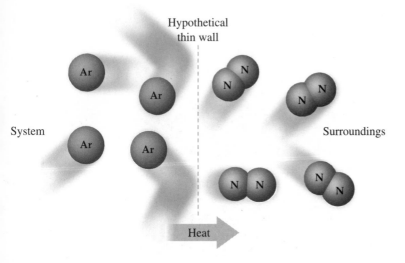

FIGURE 6.5   Heat transfer between a system and surroundings

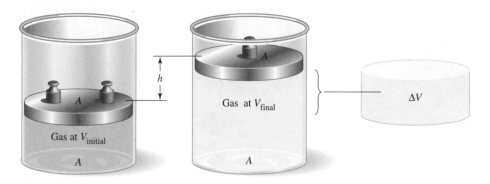

FIGURE 6.8   Pressure-volume work

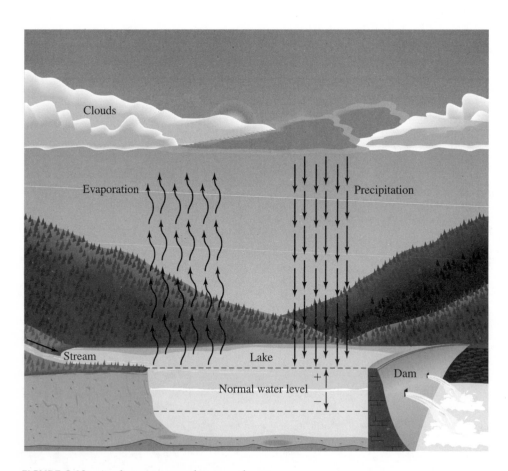

FIGURE 6.10   Analogy to internal energy changes

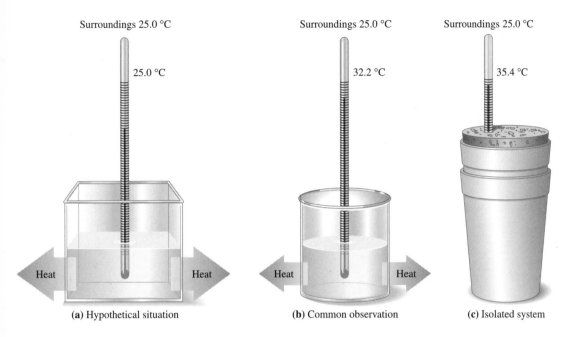

Surroundings 25.0 °C   Surroundings 25.0 °C   Surroundings 25.0 °C

25.0 °C   32.2 °C   35.4 °C

Heat   Heat   Heat   Heat

(a) Hypothetical situation   **(b)** Common observation   **(c)** Isolated system

FIGURE 6.11   Conceptualizing an exothermic reaction

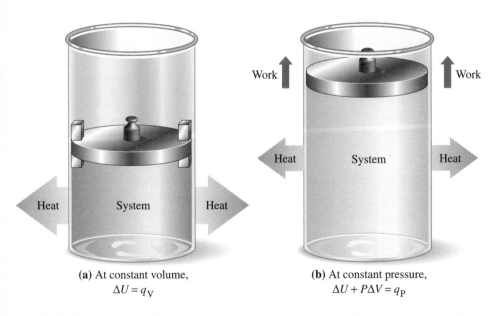

Work   Work

Heat   System   Heat   Heat   System   Heat

(a) At constant volume,   **(b)** At constant pressure,
$\Delta U = q_V$ $\Delta U + P\Delta V = q_P$

FIGURE 6.12   Comparing heats of reaction at constant volume and constant pressure for an exothermic reaction

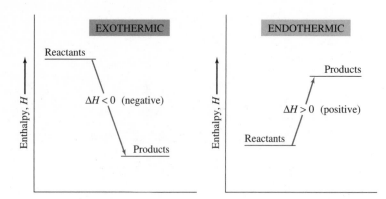

FIGURE 6.14 Enthalpy diagrams

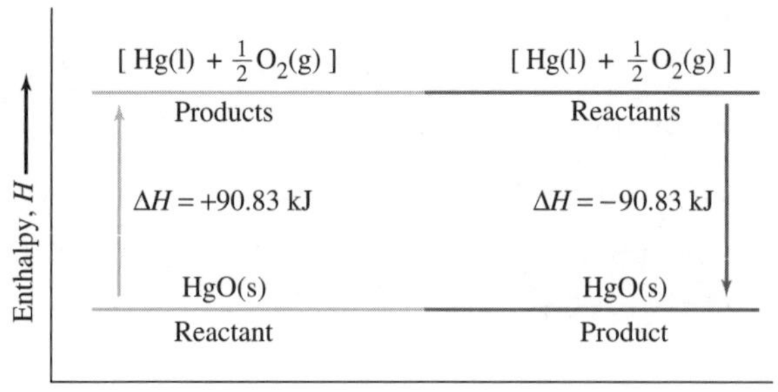

FIGURE 6.15 Reversing a chemical reaction

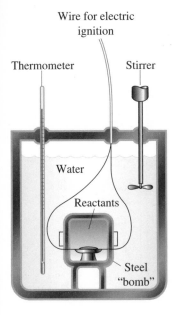

FIGURE 6.18   A bomb calorimeter

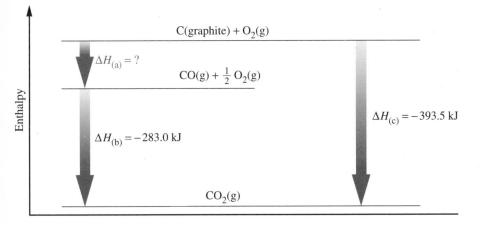

FIGURE 6.19   Determining an unknown enthalpy change through an enthalpy diagram

Chapter 7

Atomic Structure

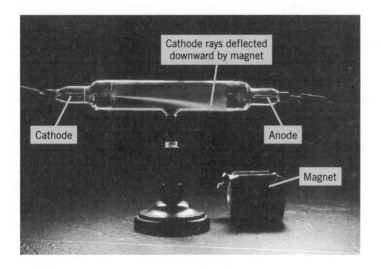

FIGURE 7.1   Cathode rays and their deflection in a magnetic field

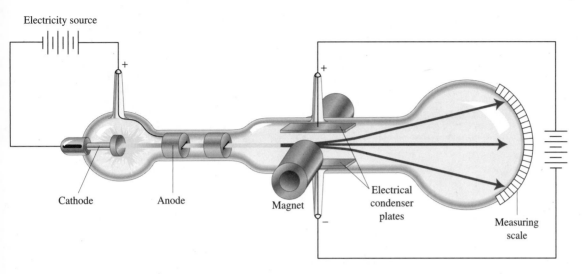

FIGURE 7.2   Thomson's apparatus for determining the mass-to-charge ratio, m_e/e, of cathode rays

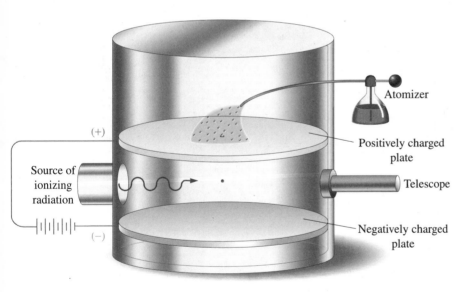

FIGURE 7.3   Millikan's oil-drop experiment

(a) The observations

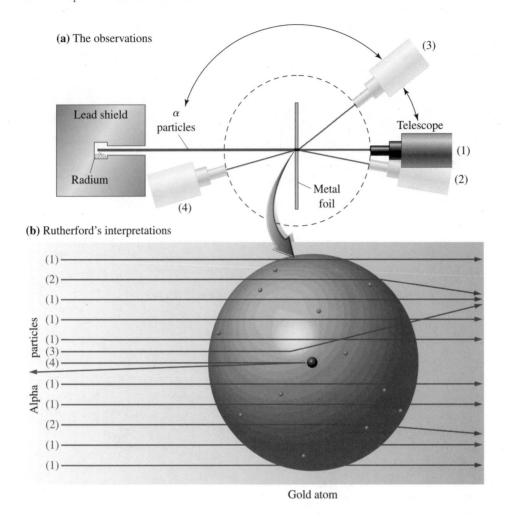

(b) Rutherford's interpretations

FIGURE 7.5   The scattering of alpha (α) particles by a thin metal foil

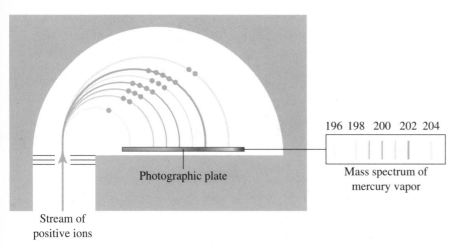

196 198 200 202 204

Mass spectrum of
mercury vapor

Photographic plate

Stream of
positive ions

FIGURE 7.6   A mass spectrometer

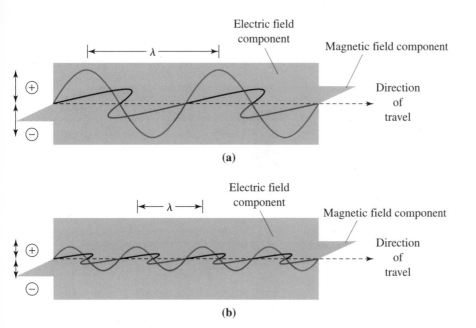

FIGURE 7.9   An electromagnetic wave

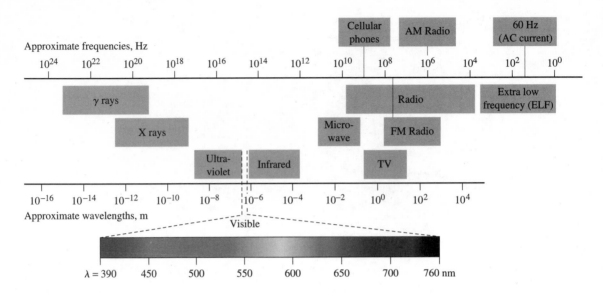

FIGURE 7.10   The electromagnetic spectrum

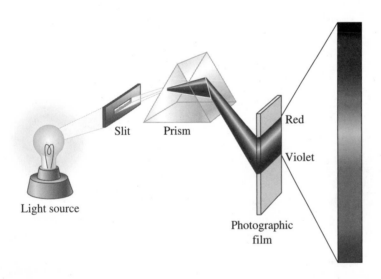

FIGURE 7.11   The spectrum of white light

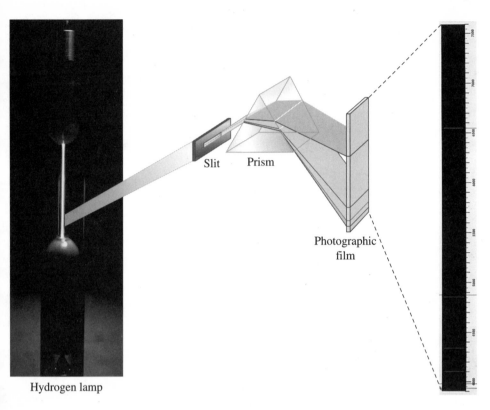

FIGURE 7.12   The emission spectrum of hydrogen

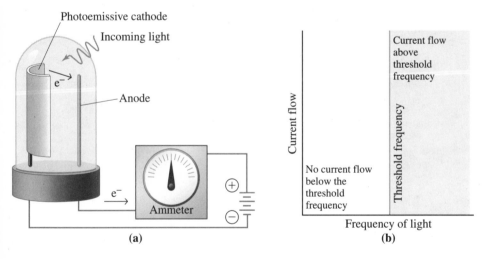

FIGURE 7.15   The photoelectric effect and the frequency of light

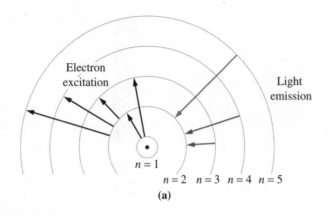

(a)

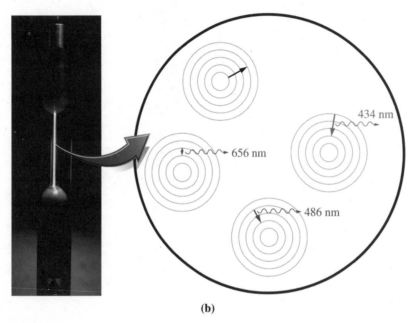

(b)

FIGURE 7.17   The Bohr model of the hydrogen atom

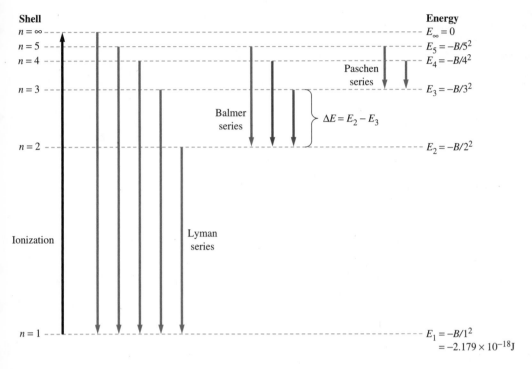

FIGURE 7.18   Energy levels and spectral lines for hydrogen

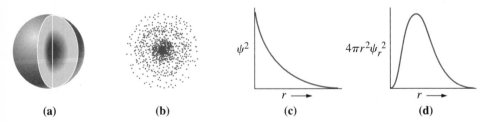

(a)   (b)   (c)   (d)

FIGURE 7.23   The 1s orbital

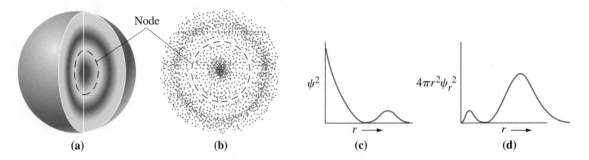

FIGURE 7.25   The 2*s* orbital

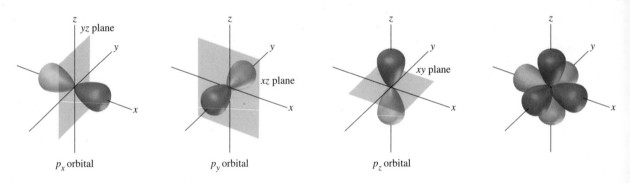

FIGURE 7.26   The three 2*p* orbitals

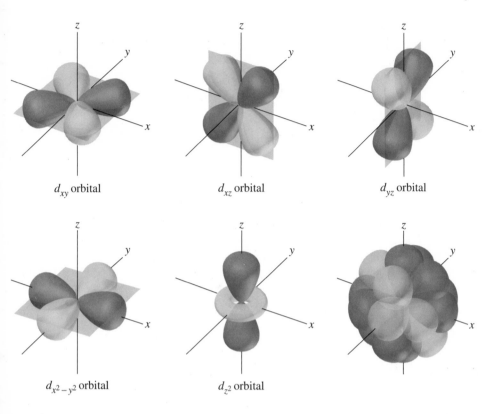

d_{xy} orbital d_{xz} orbital d_{yz} orbital

$d_{x^2-y^2}$ orbital d_{z^2} orbital

FIGURE 7.27   The five d orbitals

Chapter 8

Electron Configurations, Atomic Properties, and the Periodic Table

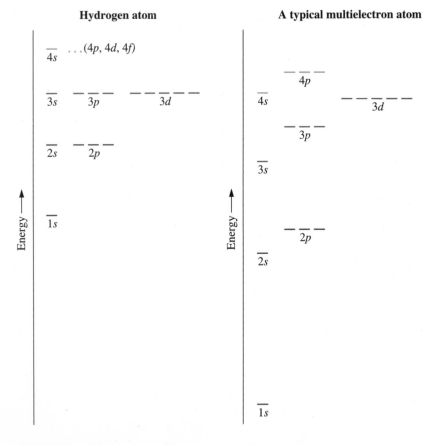

FIGURE 8.1 Orbital energy diagrams

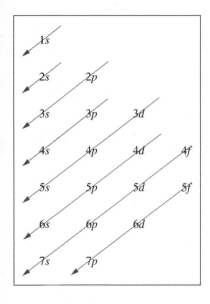

FIGURE 8.2   The order in which subshells are filled with electrons

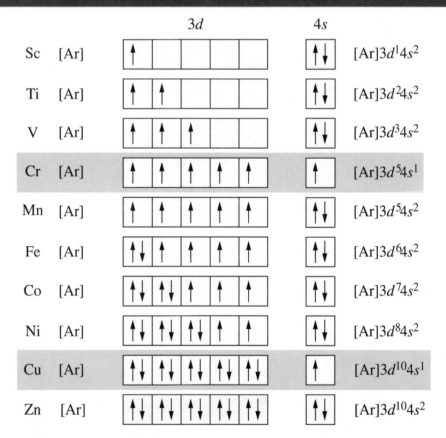

TABLE 8.2   Electron Configurations of the First Transition Series Elements

1A												3A	4A	5A	6A	7A	8A
1 **H** $1s^1$	2A																2 **He** $1s^2$
3 **Li** $2s^1$	4 **Be** $2s^2$	3B	4B	5B	6B	7B		8B		1B	2B	5 **B** $2s^2 2p^1$	6 **C** $2s^2 2p^2$	7 **N** $2s^2 2p^3$	8 **O** $2s^2 2p^4$	9 **F** $2s^2 2p^5$	10 **Ne** $2s^2 2p^6$
11 **Na** $3s^1$	12 **Mg** $3s^2$											13 **Al** $3s^2 3p^1$	14 **Si** $3s^2 3p^2$	15 **P** $3s^2 3p^3$	16 **S** $3s^2 3p^4$	17 **Cl** $3s^2 3p^5$	18 **Ar** $3s^2 3p^6$
19 **K** $4s^1$	20 **Ca** $4s^2$	21 **Sc** $3d^1 4s^2$	22 **Ti** $3d^2 4s^2$	23 **V** $3d^3 4s^2$	24 **Cr** $3d^5 4s^1$	25 **Mn** $3d^5 4s^2$	26 **Fe** $3d^6 4s^2$	27 **Co** $3d^7 4s^2$	28 **Ni** $3d^8 4s^2$	29 **Cu** $3d^{10} 4s^1$	30 **Zn** $3d^{10} 4s^2$	31 **Ga** $4s^2 4p^1$	32 **Ge** $4s^2 4p^2$	33 **As** $4s^2 4p^3$	34 **Se** $4s^2 4p^4$	35 **Br** $4s^2 4p^5$	36 **Kr** $4s^2 4p^6$
37 **Rb** $5s^1$	38 **Sr** $5s^2$	39 **Y** $4d^1 5s^2$	40 **Zr** $4d^2 5s^2$	41 **Nb** $4d^4 5s^1$	42 **Mo** $4d^5 5s^1$	43 **Tc** $4d^5 5s^2$	44 **Ru** $4d^7 5s^1$	45 **Rh** $4d^8 5s^1$	46 **Pd** $4d^{10}$	47 **Ag** $4d^{10} 5s^1$	48 **Cd** $4d^{10} 5s^2$	49 **In** $5s^2 5p^1$	50 **Sn** $5s^2 5p^2$	51 **Sb** $5s^2 5p^3$	52 **Te** $5s^2 5p^4$	53 **I** $5s^2 5p^5$	54 **Xe** $5s^2 5p^6$
55 **Cs** $6s^1$	56 **Ba** $6s^2$	57 *La $5d^1 6s^2$	72 **Hf** $5d^2 6s^2$	73 **Ta** $5d^3 6s^2$	74 **W** $5d^4 6s^2$	75 **Re** $5d^5 6s^2$	76 **Os** $5d^6 6s^2$	77 **Ir** $5d^7 6s^2$	78 **Pt** $5d^9 6s^1$	79 **Au** $5d^{10} 6s^1$	80 **Hg** $5d^{10} 6s^2$	81 **Tl** $6s^2 6p^1$	82 **Pb** $6s^2 6p^2$	83 **Bi** $6s^2 6p^3$	84 **Po** $6s^2 6p^4$	85 **At** $6s^2 6p^5$	86 **Rn** $6s^2 6p^6$
87 **Fr** $7s^1$	88 **Ra** $7s^2$	89 †Ac $6d^1 7s^2$	104 **Rf** $6d^2 7s^2$	105 **Db** $6d^3 7s^2$	106 **Sg** $6d^4 7s^2$	107 **Bh**	108 **Hs**	109 **Mt**	110 **Ds**	111	112	Unknown	114	Unknown	116		

*	58 **Ce** $4f^2 6s^2$	59 **Pr** $4f^3 6s^2$	60 **Nd** $4f^4 6s^2$	61 **Pm** $4f^5 6s^2$	62 **Sm** $4f^6 6s^2$	63 **Eu** $4f^7 6s^2$	64 **Gd** $4f^7 5d^1 6s^2$	65 **Tb** $4f^9 6s^2$	66 **Dy** $4f^{10} 6s^2$	67 **Ho** $4f^{11} 6s^2$	68 **Er** $4f^{12} 6s^2$	69 **Tm** $4f^{13} 6s^2$	70 **Yb** $4f^{14} 6s^2$	71 **Lu** $4f^{14} 5d^1 6s^2$
†	90 **Th** $6d^2 7s^2$	91 **Pa** $5f^2 6d^1 7s^2$	92 **U** $5f^3 6d^1 7s^2$	93 **Np** $5f^4 6d^1 7s^2$	94 **Pu** $5f^6 7s^2$	95 **Am** $5f^7 7s^2$	96 **Cm** $5f^7 6d^1 7s^2$	97 **Bk** $5f^9 7s^2$	98 **Cf** $5f^{10} 7s^2$	99 **Es** $5f^{11} 7s^2$	100 **Fm** $5f^{12} 7s^2$	101 **Md** $5f^{13} 7s^2$	102 **No** $5f^{14} 7s^2$	103 **Lr** $5f^{14} 6d^1 7s^2$

FIGURE 8.3   Electron configurations and the periodic table

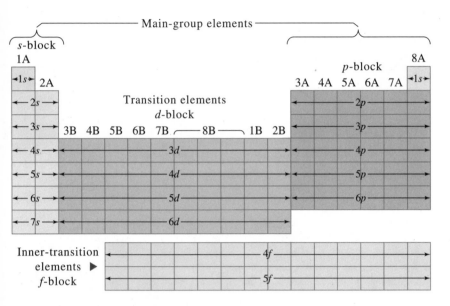

FIGURE 8.6   The periodic table and the order of filling of subshells

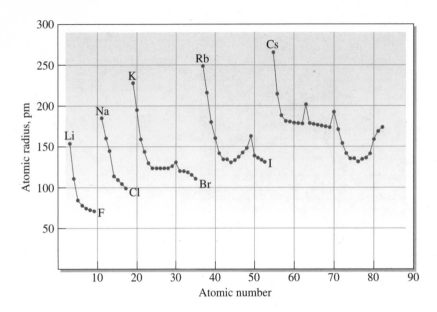

FIGURE 8.9   Atomic radii of the elements

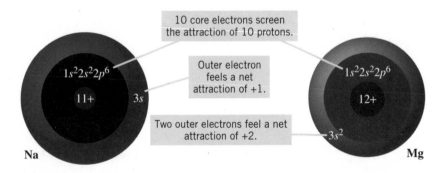

FIGURE 8.10   A simplified view of shielding and effective nuclear charge

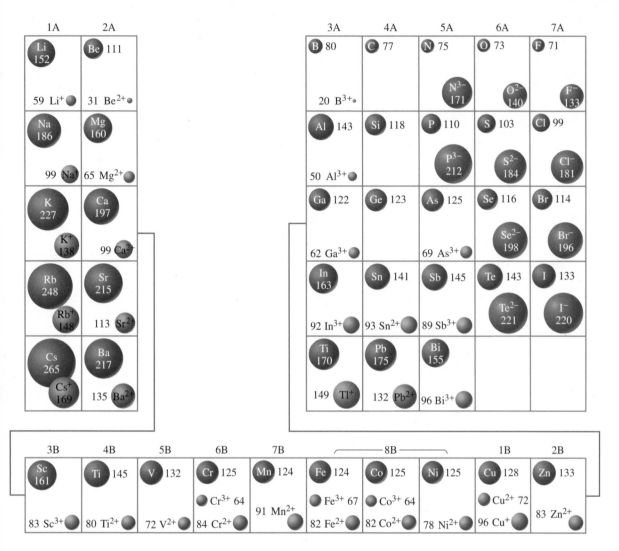

FIGURE 8.14   Some representative atomic and ionic radii

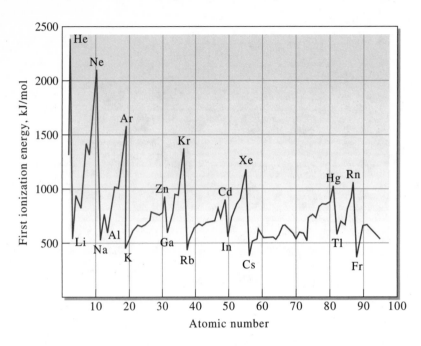

FIGURE 8.15 First ionization energy as a function of atomic number

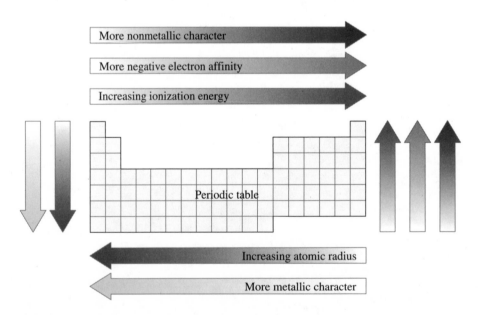

FIGURE 8.17 Atomic properties–A summary of trends in the periodic table

Chapter 9

Chemical Bonds

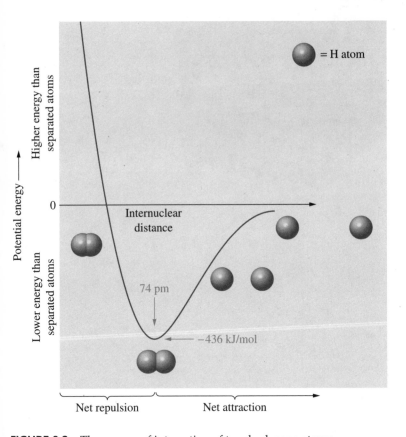

FIGURE 9.2 The energy of interaction of two hydrogen atoms

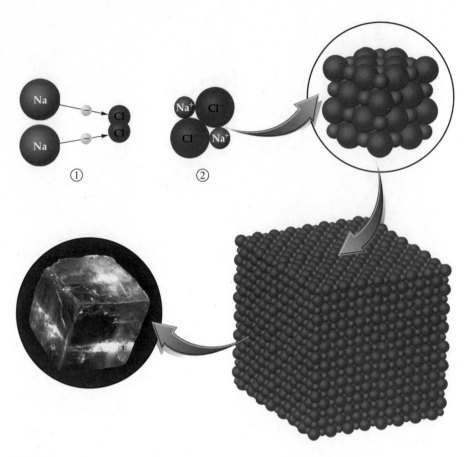

FIGURE 9.4 Formation of a crystal of sodium chloride

FIGURE 9.5   Born-Haber cycle for 1 mol of sodium chloride

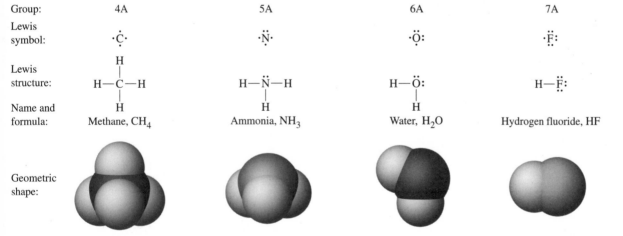

Group:	4A	5A	6A	7A
Lewis symbol:	$\cdot\overset{\displaystyle\cdot}{\underset{}{C}}\cdot$	$\cdot\overset{\displaystyle\cdot\cdot}{\underset{\displaystyle\cdot}{N}}\cdot$	$\cdot\overset{\displaystyle\cdot\cdot}{\underset{\displaystyle\cdot\cdot}{O}}:$	$\cdot\overset{\displaystyle\cdot\cdot}{\underset{\displaystyle\cdot\cdot}{F}}:$
Lewis structure:	H—C—H with H above and below, $H-\underset{H}{\overset{H}{C}}-H$	$H-\underset{H}{\overset{\cdot\cdot}{N}}-H$	$H-\underset{H}{\overset{\cdot\cdot}{O}}:$	$H-\overset{\cdot\cdot}{\underset{\cdot\cdot}{F}}:$
Name and formula:	Methane, CH_4	Ammonia, NH_3	Water, H_2O	Hydrogen fluoride, HF
Geometric shape:				

FIGURE 9.6   Four hydrogen compounds of second-period nonmetals

	1A	2A	3B	4B	5B	6B	7B	8B			1B	2B	3A	4A	5A	6A	7A

Below 1.0   2.0–2.4
1.0–1.4   2.5–2.9
1.5–1.9   3.0–4.0

Period	1A	2A	3B	4B	5B	6B	7B	8B			1B	2B	3A	4A	5A	6A	7A
1	**H** 2.1																
2	**Li** 1.0	**Be** 1.5											**B** 2.0	**C** 2.5	**N** 3.0	**O** 3.5	**F** 4.0
3	**Na** 0.9	**Mg** 1.2											**Al** 1.5	**Si** 1.8	**P** 2.1	**S** 2.5	**Cl** 3.0
4	**K** 0.8	**Ca** 1.0	**Sc** 1.3	**Ti** 1.5	**V** 1.6	**Cr** 1.6	**Mn** 1.5	**Fe** 1.8	**Co** 1.8	**Ni** 1.8	**Cu** 1.9	**Zn** 1.7	**Ga** 1.6	**Ge** 1.8	**As** 2.0	**Se** 2.4	**Br** 2.8
5	**Rb** 0.8	**Sr** 1.0	**Y** 1.2	**Zr** 1.4	**Nb** 1.6	**Mo** 1.8	**Tc** 1.9	**Ru** 2.2	**Rh** 2.2	**Pd** 2.2	**Ag** 1.9	**Cd** 1.7	**In** 1.7	**Sn** 1.8	**Sb** 1.9	**Te** 2.1	**I** 2.5
6	**Cs** 0.7	**Ba** 0.9	**La*** 1.1	**Hf** 1.3	**Ta** 1.5	**W** 1.7	**Re** 1.9	**Os** 2.2	**Ir** 2.2	**Pt** 2.2	**Au** 2.4	**Hg** 1.9	**Tl** 1.8	**Pb** 1.8	**Bi** 1.9	**Po** 2.0	**At** 2.2
7	**Fr** 0.7	**Ra** 0.9	**Ac†** 1.1														

*Lanthanides: 1.1–1.3
†Actinides: 1.3–1.5

FIGURE 9.8   Pauling's electronegativities of the elements

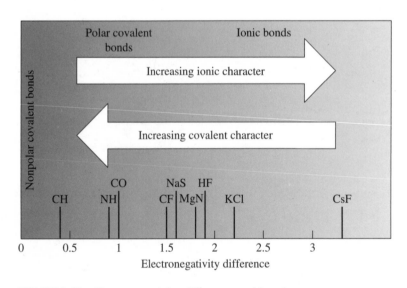

FIGURE 9.10   Electronegativity difference and bond type

Nonpolar covalent
bond

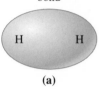

(a)

Polar covalent
bond

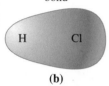

(b)

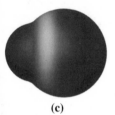

(c)   **FIGURE 9.11**   Charge distribution in nonpolar and polar covalent bonds

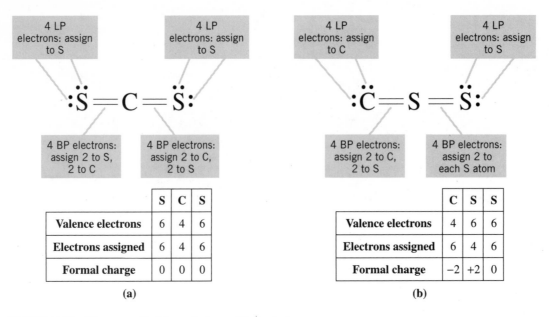

FIGURE 9.12 The concept of formal charge illustrated

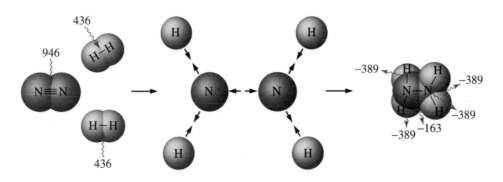

FIGURE 9.14 Bond breakage and bond formation in a reaction

Chapter 10

Bonding Theory and Molecular Structure

Table 10.1 (Part 1) VSEPR Notation, Electron-Group Geometry, and Molecular Geometry							
Number of Electron Groups	Electron-Group Geometry	Number of Lone Pairs	VSEPR Notation	Molecular Geometry	Ideal Bond Angles	Example	Molecular Model
2	Linear	0	AX_2	X—A—X Linear	180°	$BeCl_2$	
3	Trigonal planar	0	AX_3	X—A⟨X X Trigonal planar	120°	BF_3	
3	Trigonal planar	1	AX_2E	X—A X Angular	120°	SO_2	

TABLE 10.1   VSEPR Notations, Electron-Group Geometry, and Molecular Geometry

Table 10.1 (Part 2) VSEPR Notation, Electron-Group Geometry, and Molecular Geometry

Number of Electron Groups	Electron-Group Geometry	Number of Lone Pairs	VSEPR Notation	Molecular Geometry	Ideal Bond Angles	Example	Molecular Model
4	Tetrahedral	0	AX_4	Tetrahedral	109.5°	CH_4	
4	Tetrahedral	1	AX_3E	Trigonal pyramidal	109.5°	NH_3	
4	Tetrahedral	2	AX_2E_2	Angular	109.5°	OH_2	
5	Trigonal bipyramidal	0	AX_5	Trigonal bipyramidal	90°, 120°, 180°	PCl_5	

TABLE 10.1 VSEPR Notations, Electron-Group Geometry, and Molecular Geometry

Table 10.1 (Part 4)   VSEPR Notation, Electron-Group Geometry, and Molecular Geometry

Number of Electron Groups	Electron-Group Geometry	Number of Lone Pairs	VSEPR Notation	Molecular Geometry	Ideal Bond Angles	Example	Molecular Model
5	Trigonal bipyramidal	1	AX_4E	Seesaw	90°, 120°, 180°	SF_4	
5	Trigonal bipyramidal	2	AX_3E_2	T-shaped	90°, 180°	ClF_3	
5	Trigonal bipyramidal	3	AX_2E_3	Linear	180°	XeF_2	
6	Octahedral	0	AX_6	Octahedral	90°, 180°	SF_6	
6	Octahedral	1	AX_5E	Square pyramidal	90°	BrF_5	
6	Octahedral	2	AX_4E_2	Square planar	90°	XeF_4	

TABLE 10.1   VSEPR Notations, Electron-Group Geometry, and Molecular Geometry

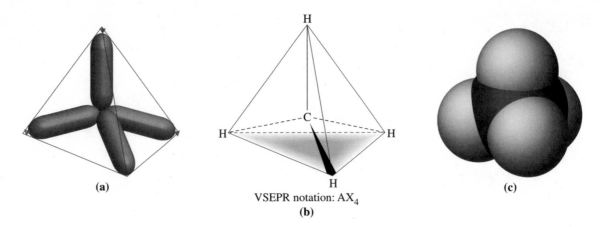

(a)

VSEPR notation: AX$_4$

(b)

(c)

FIGURE 10.3 The electron-group geometry and molecular geometry of methane

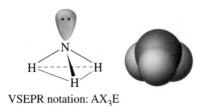

VSEPR notation: AX$_3$E

FIGURE 10.5 Molecular geometry of ammonia

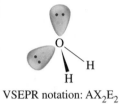

VSEPR notation: AX_2E_2 **FIGURE 10.6**   Molecular geometry of water

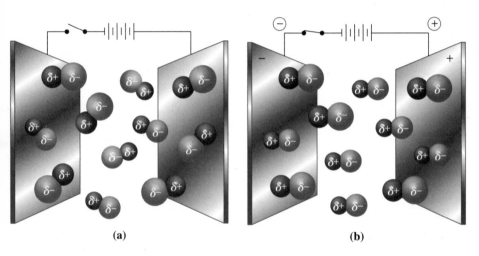

FIGURE 10.8   Behavior of polar molecules in an electric field

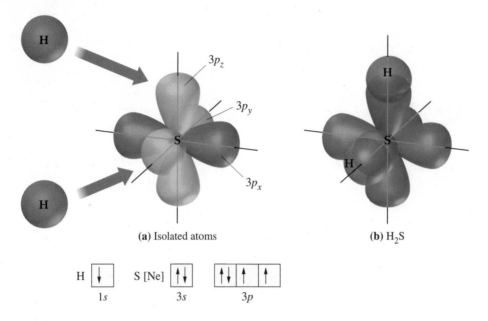

(a) Isolated atoms

(b) H_2S

FIGURE 10.11   Atomic orbital overlap and bonding in H_2S

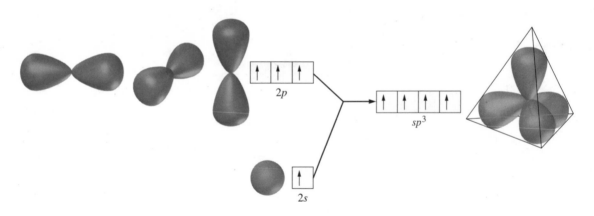

FIGURE 10.12   The sp^3 hybridization scheme for C

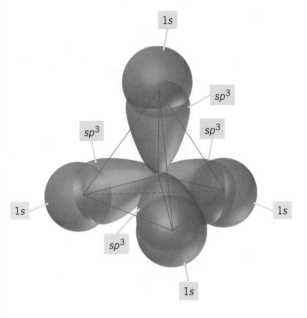

FIGURE 10.13 sp^3 hybrid orbitals and bonding in CH_4

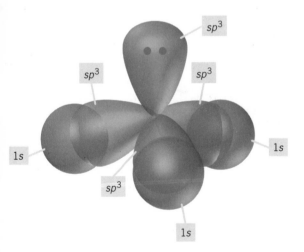

FIGURE 10.14 sp^3 hybrid orbitals and bonding in NH_3

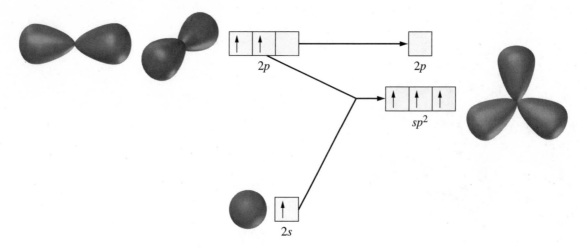

FIGURE 10.15   The sp^2 hybridization scheme for boron

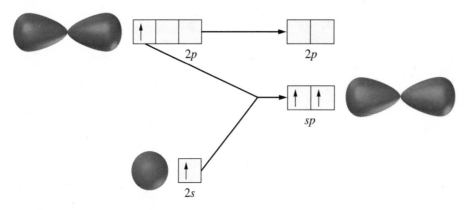

FIGURE 10.16   The sp hybridization scheme for beryllium

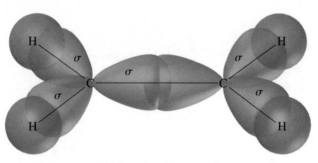

(a) The σ-bond framework

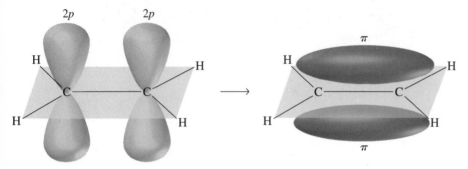

(b) The formation of a π-bond by the overlap of the half-filled $2p$ orbitals

π: C($2p$)—C($2p$)

σ: H($1s$)—C(sp^2)   H—C=C—H σ: H($1s$)—C(sp^2)
                          H   H

σ: C(sp^2)—C(sp^2)

(c) Hybridization and bonding scheme

FIGURE 10.21   Bonding in ethylene, C_2H_4, by the valence bond theory

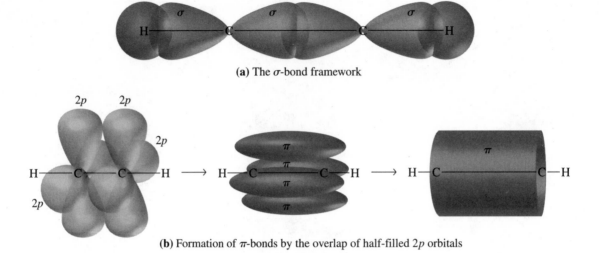

(a) The σ-bond framework

(b) Formation of π-bonds by the overlap of half-filled $2p$ orbitals

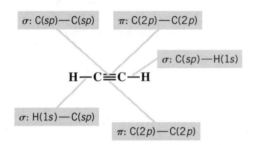

(c) Hybridization and bonding scheme

FIGURE 10.22   Bonding in acetylene, C_2H_2, by the valence bond theory

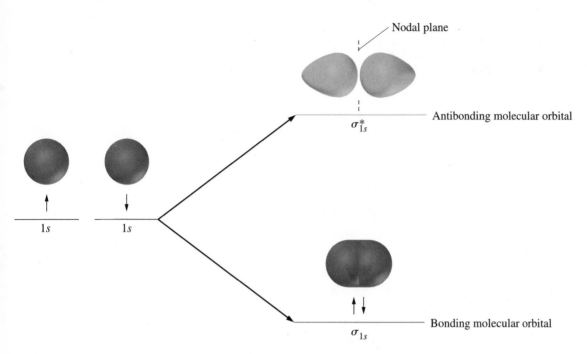

FIGURE 10.24   Molecular orbitals and bonding in the H$_2$ molecule

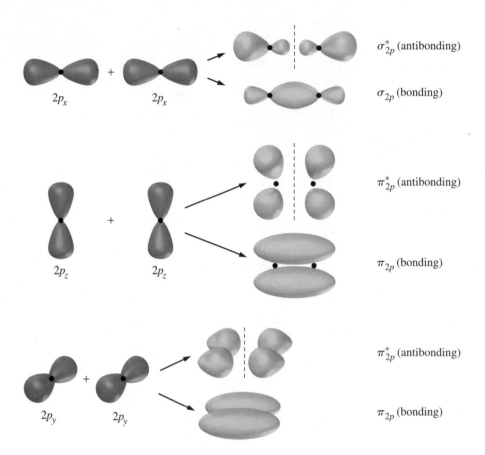

FIGURE 10.26 Molecular orbitals formed by combining $2p$ atomic orbitals

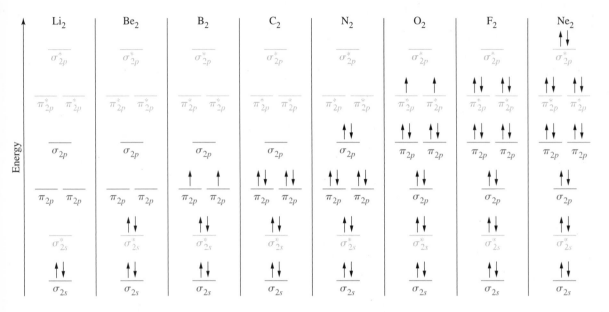

FIGURE 10.27   Relative energy levels of molecular orbitals obtained from 2s and 2p atomic orbitals, and some actual and hypothetical diatomic molecules

Chapter 11

States of Matter and Intermolecular Forces

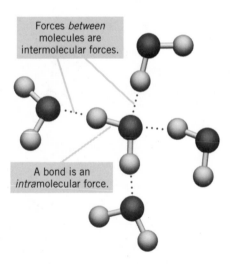

Forces *between* molecules are intermolecular forces.

A bond is an *intra*molecular force.

FIGURE 11.1   Intermolecular and intramolecular forces compared

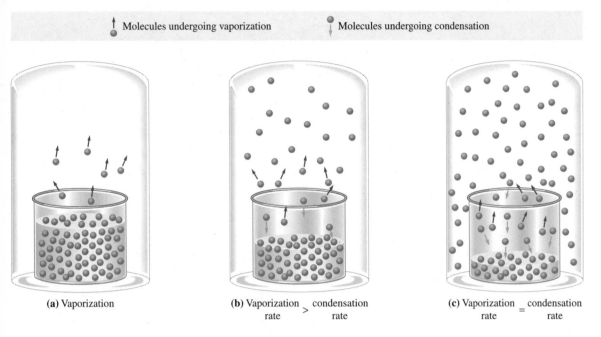

FIGURE 11.3   Liquid-vapor equilibrium and vapor pressure

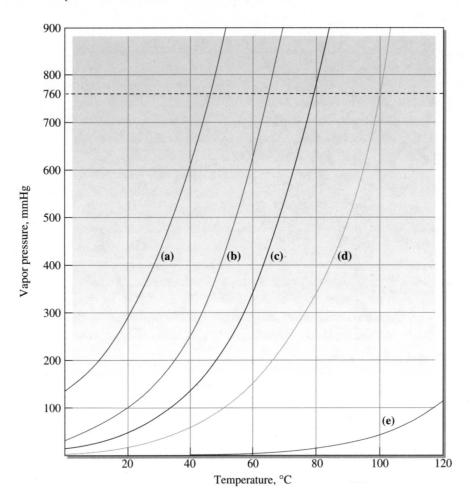

FIGURE 11.4   Vapor pressure curves of several liquids

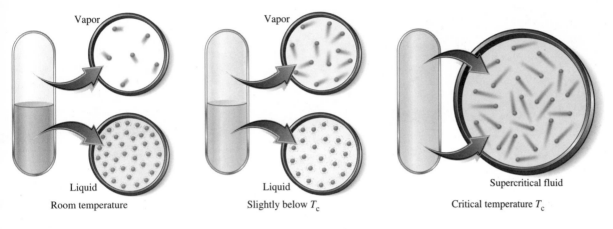

Vapor

Liquid

Room temperature

Vapor

Liquid

Slightly below T_c

Supercritical fluid

Critical temperature T_c

FIGURE 11.6   The critical point

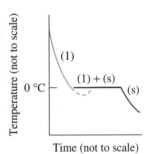

Temperature (not to scale)

$0\,°C$

(1)

(1) + (s)

(s)

Time (not to scale)

FIGURE 11.7   Cooling curve for water

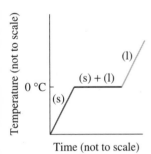

Temperature (not to scale)

$0\,°C$

(l)

(s) + (l)

(s)

Time (not to scale)

FIGURE 11.8   Heating curve for water

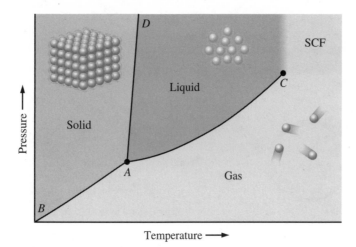

FIGURE 11.10   A generalized phase diagram: Representing temperatures, pressures, and states of matter of a substance

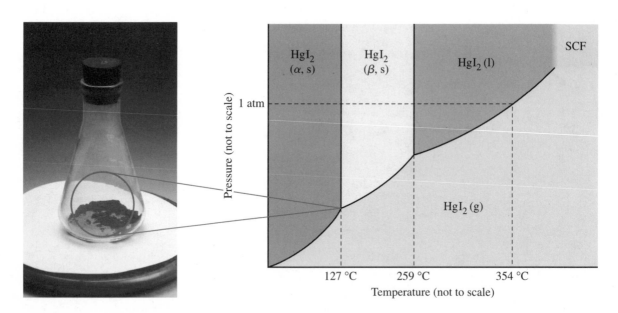

FIGURE 11.11   Phase diagram of mercury(II) iodide, HgI_2

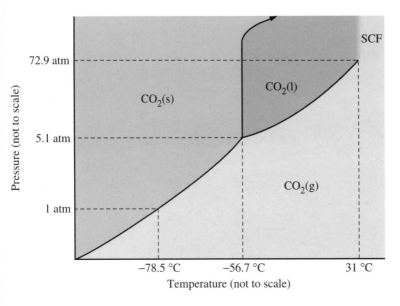

FIGURE 11.12   Phase diagram of carbon dioxide, CO_2

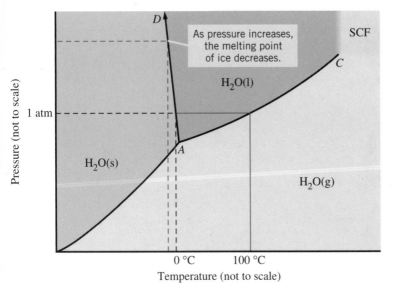

FIGURE 11.13   Phase diagram of water, H_2O

(a) Unpolarized
molecule

(b) Instantaneous
dipole

(c) Induced dipole

FIGURE 11.16   Dispersion forces

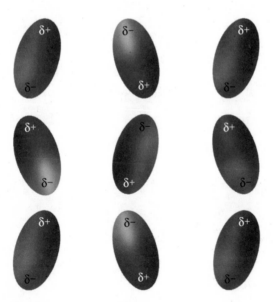

FIGURE 11.19   Dipole-dipole interactions

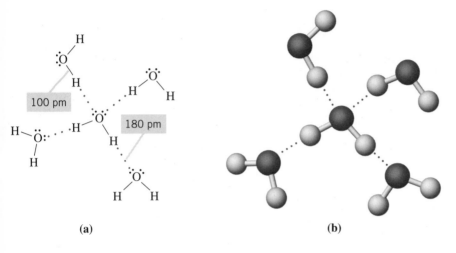

(a)   (b)

FIGURE 11.20   Hydrogen bonds in water

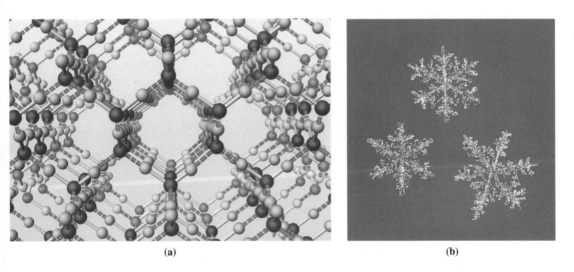

(a)   (b)

FIGURE 11.21   Hydrogen bonding in ice

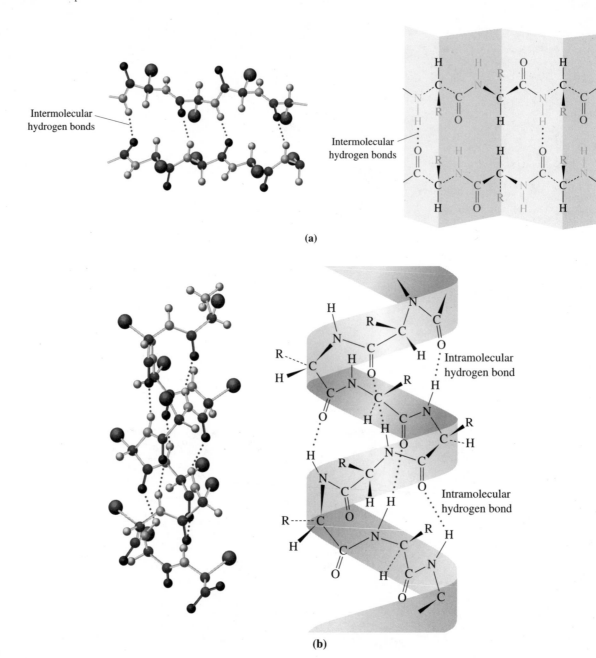

FIGURE 11.24   The two principal secondary structures of proteins

FIGURE 11.26   Intermolecular forces in a liquid

FIGURE 11.28   Adhesive and cohesive forces

Table 11.5 Some Characteristics of Crystalline Solids

Type	Structural Particles	Intermolecular Forces	Typical Properties	Examples
Molecular				
Nonpolar	Atoms or nonpolar molecules	Dispersion forces	Extremely low to moderate melting points; soluble in nonpolar solvents	Ar, H_2, I_2, CCl_4, CH_4, CO_2
Polar	Polar molecules	Dispersion forces, dipole–dipole and dipole-induced dipole attractions	Low to moderate melting points; soluble in some polar and some nonpolar solvents	HCl, H_2S, $CHCl_3$, $(CH_3)_2O$, $(CH_3)_2CO$
Hydrogen-bonded	Molecules with H bonded to N, O, or F	Hydrogen bonds	Low to moderate melting points; solublein some hydrogen-bonded and somepolar liquids	H_2O, HF, NH_3, CH_3OH, CH_3COOH
Network Covalent	Atoms	Covalent bonds	Most are very hard; sublime or melt at very high temperatures; most are nonconductors of electricity	C(diamond), C(graphite) SiC, SiO_2, BN
Ionic	Cations and anions	Electrostatic attractions	Hard; brittle; moderate to very high melting points; nonconductors as solids, but electrical conductors as liquids; many are soluble in water	NaCl, CaF_2, K_2S, MgO
Metallic	Cations and delocalized electrons	Metallic bonds	Hardness varies from soft to very hard; melting points vary from low to very high; lustrous; ductile; malleable;good to excellent conductors of heat and electricity	Na, Mg, Al, Fe, Cu, Zn, Mo, Ag, Cd, W, Pt, Hg, Pb

TABLE 11.5   Some Characteristics of Crystalline Solids

FIGURE 11.32   The crystal structure of diamond

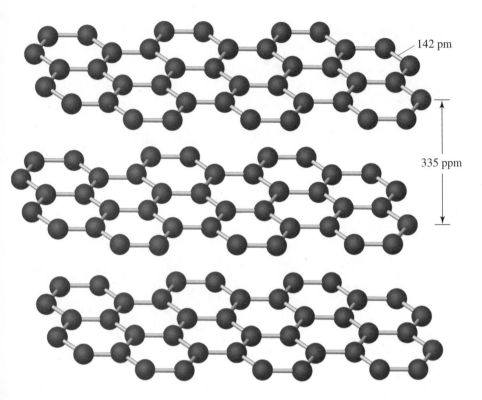

142 pm

335 ppm

FIGURE 11.33 The crystal structure of graphite

FIGURE 11.34 A ball-and-stick model of C_{60}, a buckyball

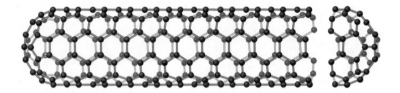

FIGURE 11.35 A ball and stick model of a carbon nanotube

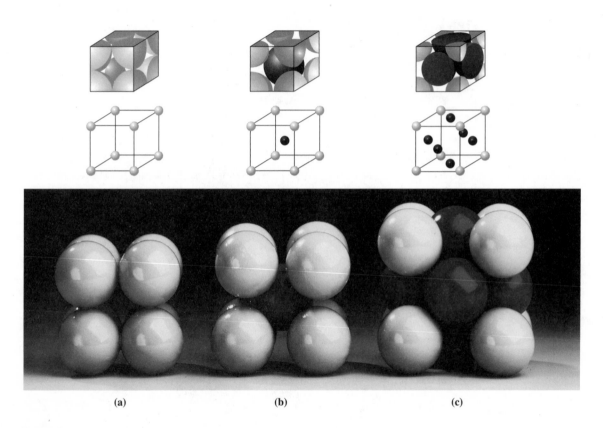

(a)   (b)   (c)

FIGURE 11.39 Unit cells in cubic crystal structures

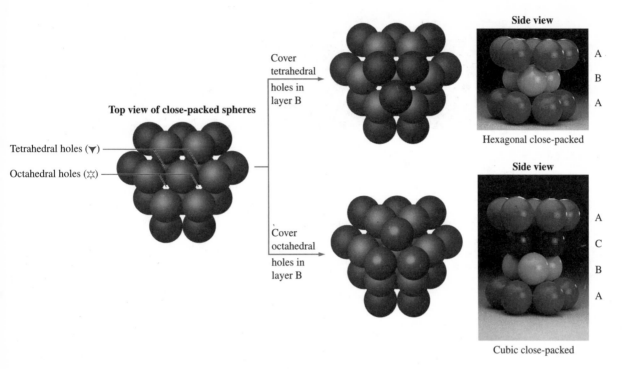

Top view of close-packed spheres

Tetrahedral holes (▼)

Octahedral holes (☆)

Cover tetrahedral holes in layer B

Cover octahedral holes in layer B

Side view

A
B
A

Hexagonal close-packed

Side view

A
C
B
A

Cubic close-packed

FIGURE 11.42   Close-packing of spheres in three dimensions

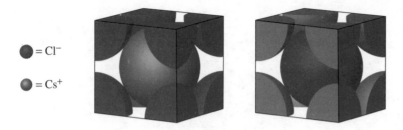

FIGURE 11.46   A unit cell of cesium chloride

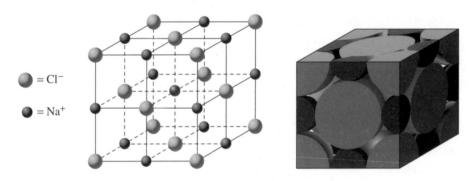

FIGURE 11.47   A unit cell of sodium chloride

Chapter 12

Physical Properties of Solutions

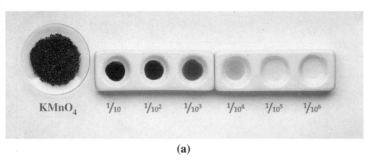

KMnO$_4$ $^1/_{10}$ $^1/_{10^2}$ $^1/_{10^3}$ $^1/_{10^4}$ $^1/_{10^5}$ $^1/_{10^6}$

(a)

(b)

FIGURE 12.1   Visualizing one part per million

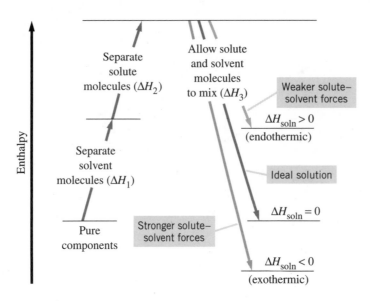

FIGURE 12.3 Visualizing an enthalpy of solution with an enthalpy diagram

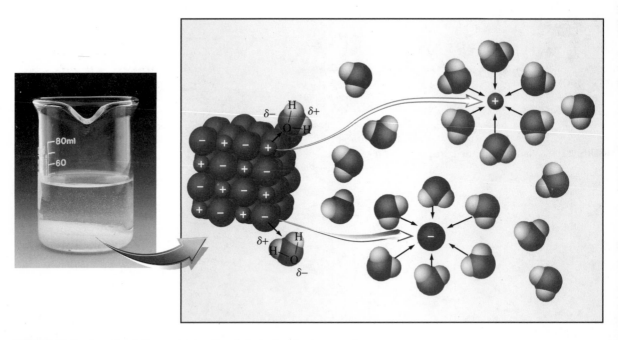

FIGURE 12.7 Ion-dipole forces in the dissolving of an ionic crystal

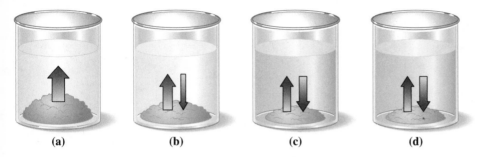

(a)   (b)   (c)   (d)

FIGURE 12.9   Formation of a saturated solution

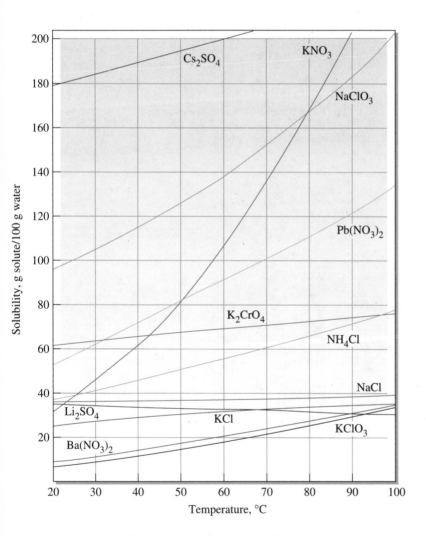

FIGURE 12.10   The solubility curves for several salts in water

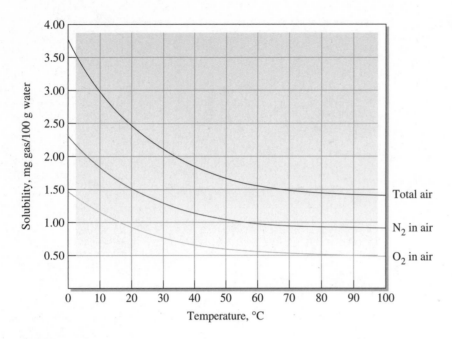

FIGURE 12.12   Solubility of air in water as a function of temperature at 1 atm pressure

FIGURE 12.13   The effect of pressure on the solubility of a gas

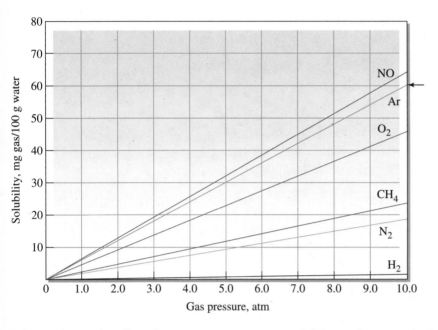

FIGURE 12.14   The effect of gas pressure on aqueous solubilities of gases at 20 °C

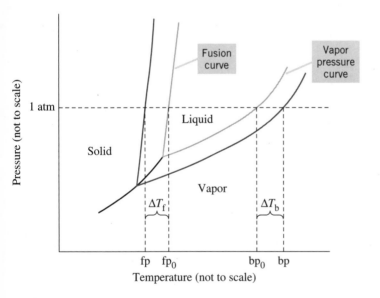

FIGURE 12.17   Vapor pressure lowering by a nonvolatile solute

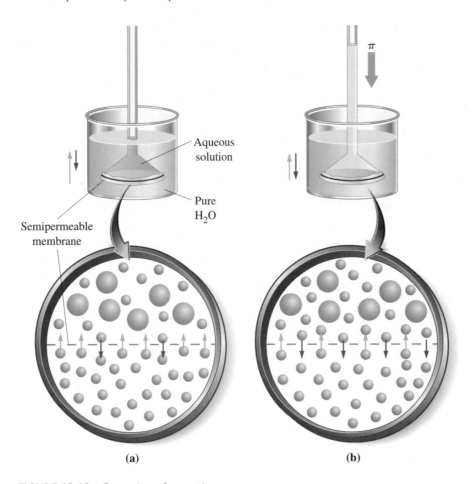

Aqueous
solution

Pure
H₂O

Semipermeable
membrane

(a)

π

(b)

FIGURE 12.18 Osmosis and osmotic pressure

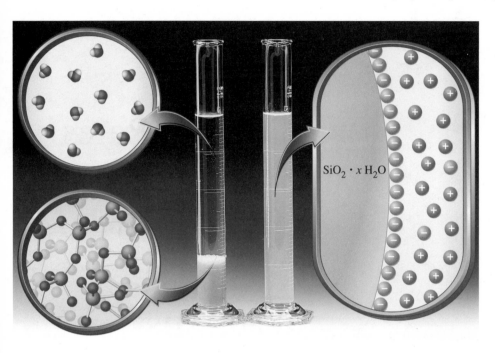

FIGURE 12.20   A suspension and a colloid

13

Chemical Kinetics: Rates and Mechanisms of Chemical Reactions

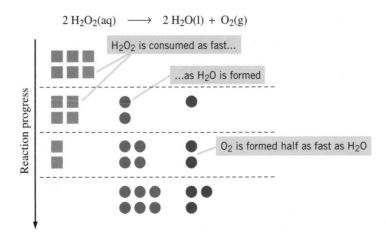

$$2\ H_2O_2(aq) \longrightarrow 2\ H_2O(l) + O_2(g)$$

H₂O₂ is consumed as fast...

...as H₂O is formed

Reaction progress

O₂ is formed half as fast as H₂O

FIGURE 13.3   Stoichiometry and rate of decomposition of H_2O_2

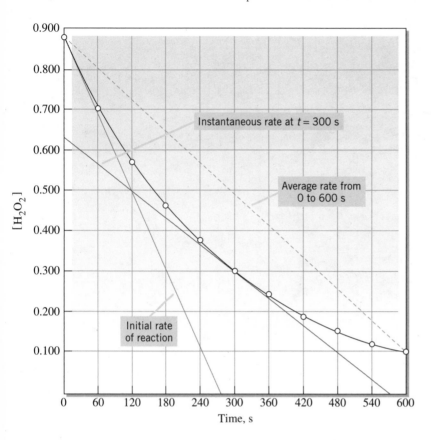

FIGURE 13.5   Kinetic data for the reaction $H_2O_2(aq) \rightarrow H_2O(I) + \dfrac{1}{2} O_2(g)$

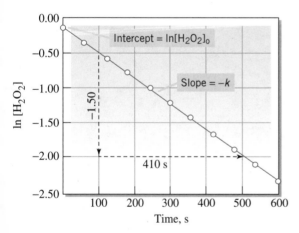

FIGURE 13.6   Test for a first-order reaction: Decomposition of H_2O_2 (aq)

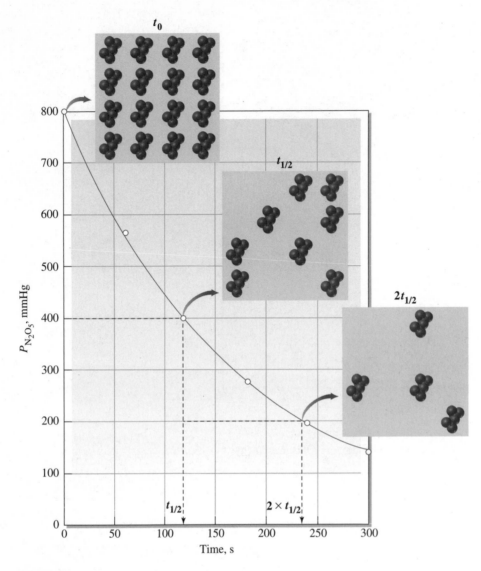

FIGURE 13.7   Decomposition of N_2O_5 at 67 °C

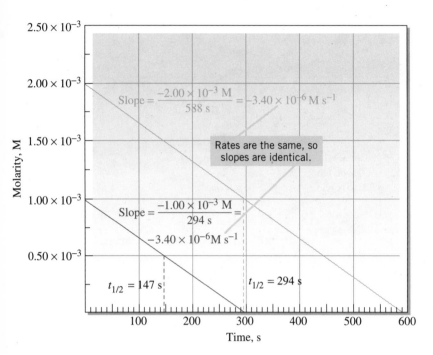

FIGURE 13.8   The decomposition of ammonia on a tungsten surface at 1100 °C: a zero-order reaction

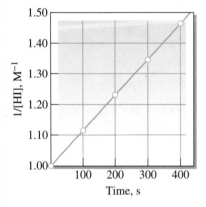

FIGURE 13.9   The decomposition of hydrogen iodide at 700 K: a second-order reaction

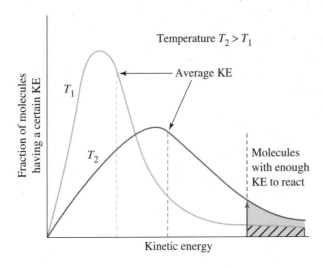

FIGURE 13.10   Distribution of kinetic energies of molecules

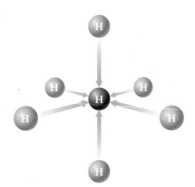

FIGURE 13.11   A reaction in which the orientation of colliding molecules is unimportant

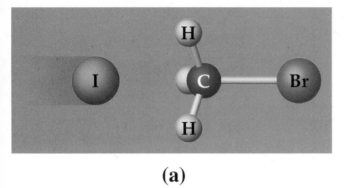

(a)

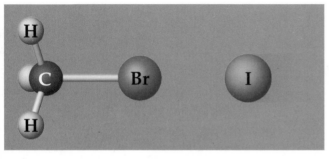

(b)

FIGURE 13.12   The importance of orientation of colliding molecules

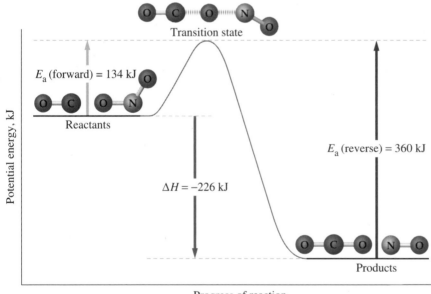

FIGURE 13.13   A reaction profile for the reaction $CO(g)+NO_2(g)\rightarrow CO_2(g)+NO(g)$

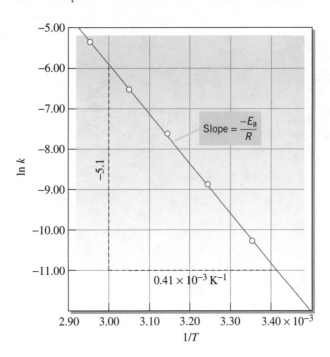

FIGURE 13.15 A plot of ln k versus 1/T for the decomposition of dinitrogen pentoxide: $N_2O_5(g) \rightarrow 2NO_2(g) + {}^1/_2 O_2(g)$

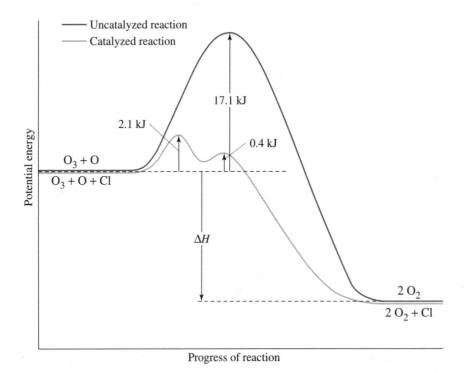

FIGURE 13.19 Reaction profile for the uncatalyzed (red) and catalyzed (blue) decomposition of ozone (Please see figure in text for actual color scheme).

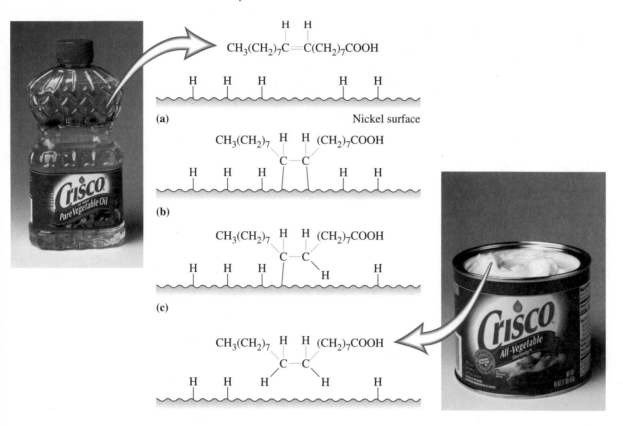

FIGURE 13.21   Heterogeneous catalysis on a nickel surface

Chapter 14

Chemical Equilibrium

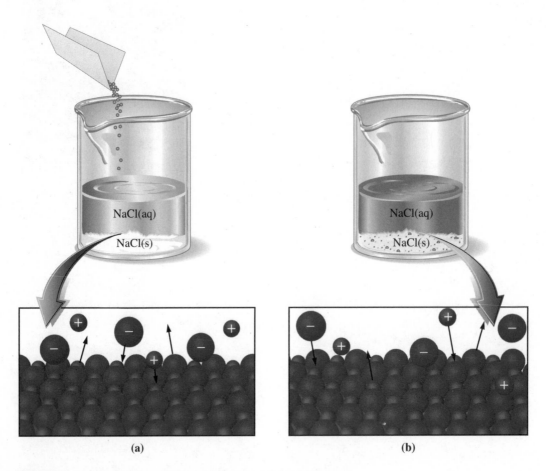

(a)   **(b)**

FIGURE 14.1 Dynamic equilibrium in saturated solution formation

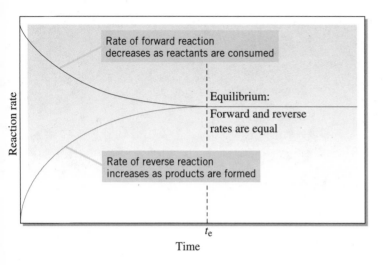

FIGURE 14.2   The concept of dynamic equilibrium

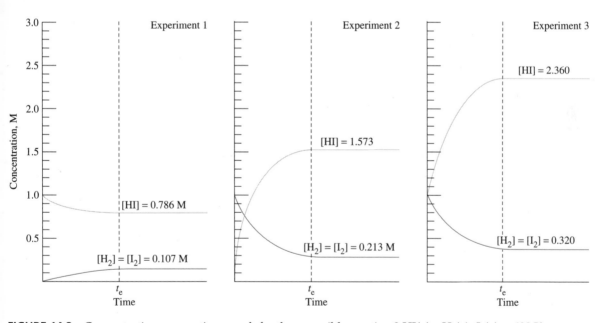

FIGURE 14.3   Concentration-versus-time graph for the reversible reaction $2\ HI(g) \rightarrow H_2(g) + I_2(g)$ at 698 K

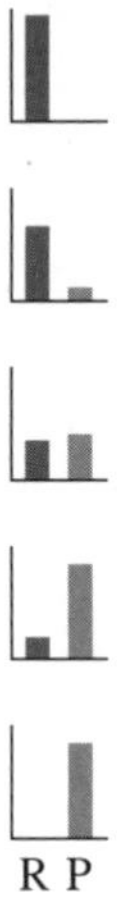

		Initial state	Net change
$Q = \dfrac{----}{\text{reactants}} = 0$		Pure reactants	$\rightarrow$ (forms products)
$Q = \dfrac{\text{products}}{\text{reactants}} < K$		Mostly reactants	$\rightarrow$ (forms products)
$Q = \dfrac{\text{products}}{\text{reactants}} = K$		At equilibrium	$\rightleftharpoons$ (none)
$Q = \dfrac{\text{products}}{\text{reactants}} > K$		Mostly products	(forms reactants) $\leftarrow$
$Q = \dfrac{\text{products}}{----} = \infty$		Pure products	(forms reactants) $\leftarrow$

R P

FIGURE 14.4   Relating Q and K and predicting the direction of net reaction

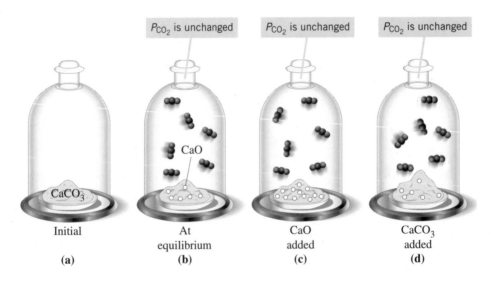

FIGURE 14.5   Equilibrium in the reaction: $CaCO_3(s) \rightleftharpoons CaO(s) + CO_2(g)$

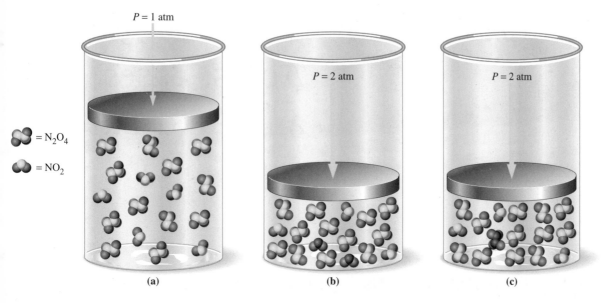

FIGURE 14.6   Illustrating Le Châtelier's principle in the reaction $2N_2O_4(g) \leftrightarrows 2NO_2(g)$ at 298 K

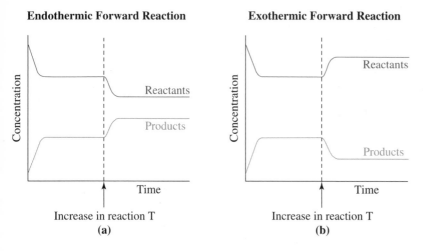

FIGURE 14.7   Effect of increasing temperature on equilibrium

Chapter 15

Acids, Bases, and Acid-Base Equilibria

Table 15.1 Relative Strengths of Some Brønsted–Lowry Acids and Their Conjugate Bases

Acid	Conjugate Base
HI (hydroiodic acid)	I^- (iodide ion)
HBr (hydrobromic acid)	Br^- (bromide ion)
HCl (hydrochloric acid)	Cl^- (chloride ion)
H_2SO_4 (sulfuric acid)	HSO_4^- (hydrogen sulfate ion)
HNO_3 (nitric acid)	NO_3^- (nitrate ion)
H_3O^+ (hydronium ion)	H_2O (water)
HSO_4^- (hydrogen sulfate ion)	SO_4^{2-} (sulfate ion)
HNO_2 (nitrous acid)	NO_2^- (nitrite ion)
HF (hydrofluoric acid)	F^- (fluoride ion)
CH_3COOH (acetic acid)	CH_3COO^- (acetate ion)
H_2CO_3 (carbonic acid)	HCO_3^- (hydrogen carbonate ion)
NH_4^+ (ammonium ion)	NH_3 (ammonia)
HCO_3^- (hydrogen carbonate ion)	CO_3^{2-} (carbonate ion)
H_2O (water)	OH^- (hydroxide ion)
CH_3OH (methanol)	CH_3O^- (methoxide ion)

Increasing acid strength

Increasing base strength

TABLE 15.1   Relative Strengths of Some Brønsted-Lowry Acids and their Conjugate Bases

Number of terminal O
atoms (red) increases   0   1   2   3

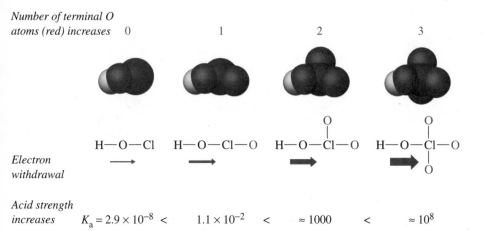

$$H-O-Cl \quad H-O-Cl-O \quad H-O-\overset{\overset{O}{|}}{Cl}-O \quad H-O-\overset{\overset{O}{|}}{\underset{\underset{O}{|}}{Cl}}-O$$

Electron
withdrawal

Acid strength
increases $K_a = 2.9 \times 10^{-8}$ < 1.1×10^{-2} < ≈ 1000 < $\approx 10^8$

FIGURE 15.2   Electron withdrawal and acid strength

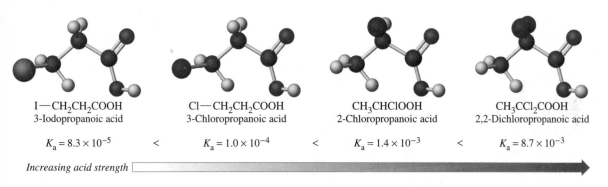

I—CH$_2$CH$_2$COOH   Cl—CH$_2$CH$_2$COOH   CH$_3$CHClOOH   CH$_3$CCl$_2$COOH
3-Iodopropanoic acid   3-Chloropropanoic acid   2-Chloropropanoic acid   2,2-Dichloropropanoic acid

$K_a = 8.3 \times 10^{-5}$ < $K_a = 1.0 \times 10^{-4}$ < $K_a = 1.4 \times 10^{-3}$ < $K_a = 8.7 \times 10^{-3}$

Increasing acid strength

FIGURE 15.3   Substituents and carboxylic acid strength

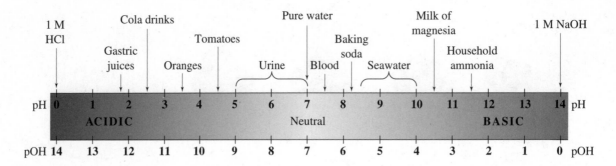

FIGURE 15.4 The pH scale

FIGURE 15.9 The common ion effect

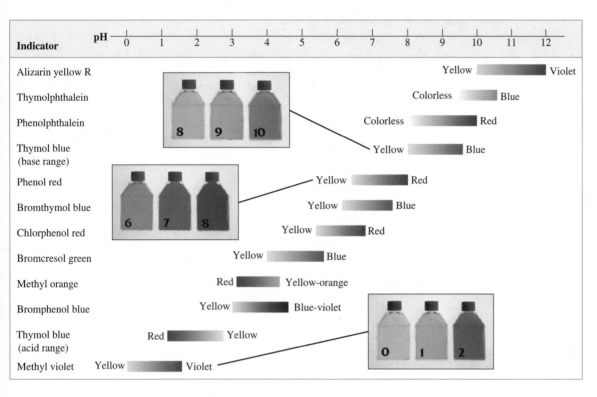

FIGURE 15.13   pH ranges and colors of several common indicators

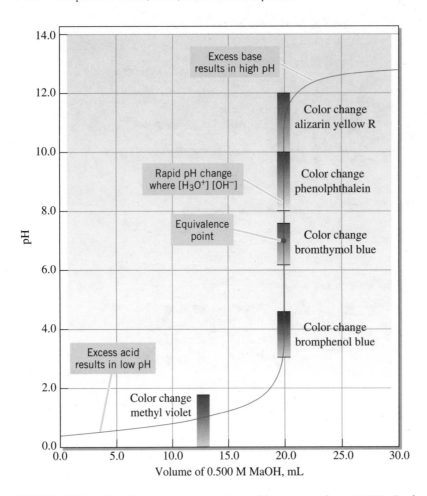

FIGURE 15.15 Titration curve for a strong acid by a strong base: 20.00 mL of 0.500 M HCl by 0.500 M NaOH

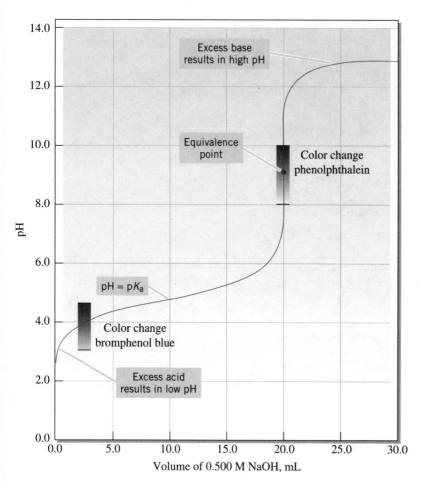

FIGURE 15.16   Titration curve for a weak acid by a strong base: 20.00 mL of
0.500 M CH_3COOH by 0.500 M NaOH

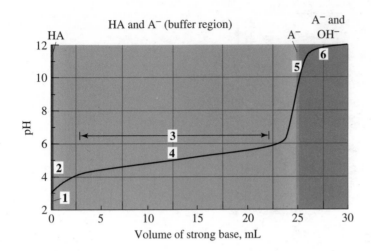

FIGURE 15.17   Summarizing types of equilibrium calculations in the titration curve for a weak acid with a strong base

Chapter 16

More Equilibria in Aqueous Solutions: Slightly Soluble Salts Complex Ions

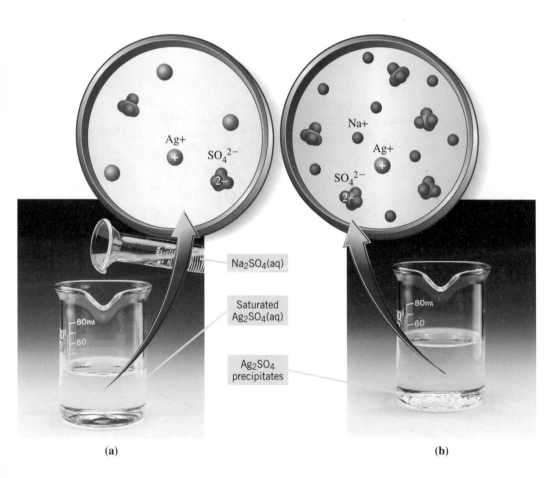

(a)   (b)

FIGURE 16.2   The common-ion effect in solubility equilibrium

FIGURE 16.8 Complex-ion formation

FIGURE 16.9 Complex-ion formation and solute solubility

FIGURE 16.12   Amphoteric behavior of AI(OH)$_3$(s)

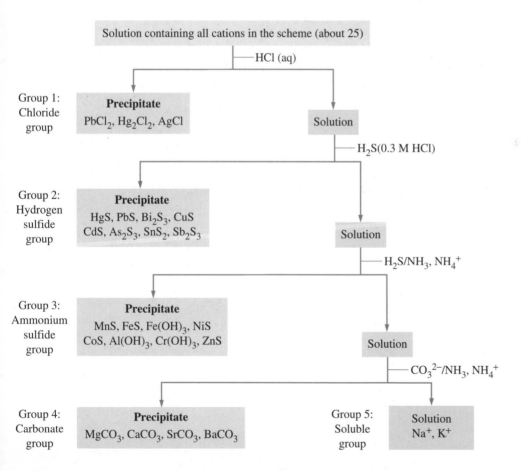

FIGURE 16.13   Outline of the qualitative analysis scheme for some common cations

Chapter 17

Thermodynamics: Spontaneity, Entropy, and Free Energy

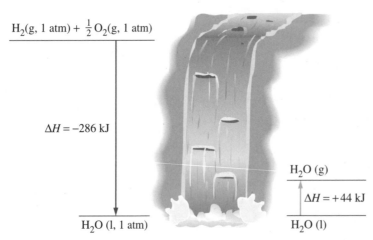

$H_2(g, 1\ atm) + \frac{1}{2} O_2(g, 1\ atm)$

$\Delta H = -286\ kJ$

$H_2O\ (l, 1\ atm)$

$H_2O\ (g)$

$\Delta H = +44\ kJ$

$H_2O\ (l)$

The formation of water at 25 °C
and 1 atm: a spontaneous process
that is exothermic

The vaporization of water at 25 °C and
pressures up to 0.0313 atm: a spontaneous
process that is endothermic

FIGURE 17.1   The direction of decrease in energy: A criterion for spontaneous change?

FIGURE 17.2   The spontaneous mixing of gases

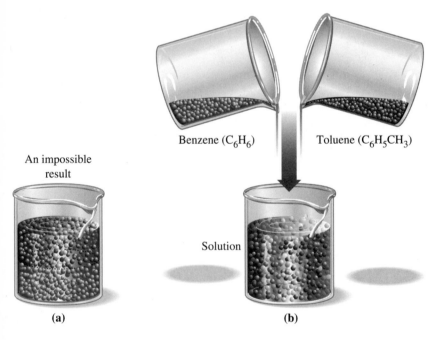

FIGURE 17.3   Formation of an ideal solution

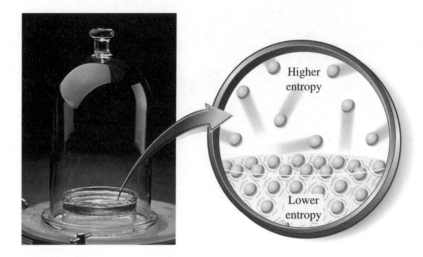

FIGURE 17.4 Increase of entropy in the vaporization of water

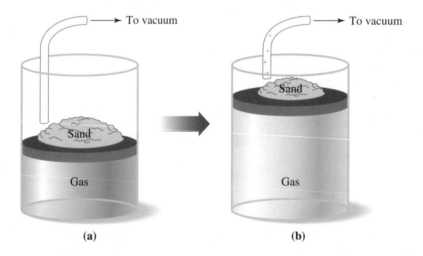

FIGURE 17.5 A nearly reversible process

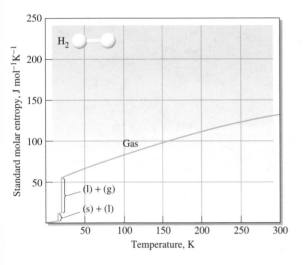

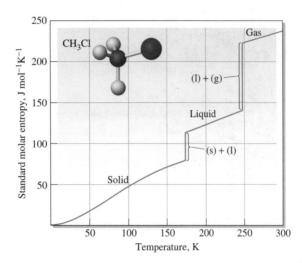

FIGURE 17.7   Entropy as a function of temperature

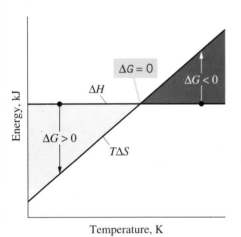

FIGURE 17.8   Gas as a criterion for spontaneous change

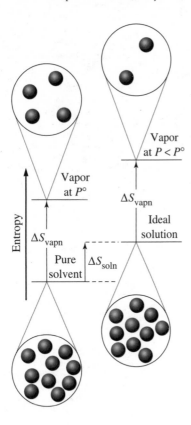

FIGURE 17.10   An entropy-based explanation of Raoult's law

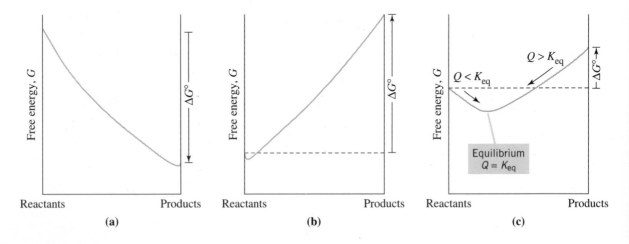

FIGURE 17.11 $\Delta G°$ and the direction and extent of spontaneous change

Chapter 18

Electrochemistry

(a)　　　　　　(b)

FIGURE 18.1 The displacement of Ag+(aq) by Cu(s): An oxidation-reduction reaction

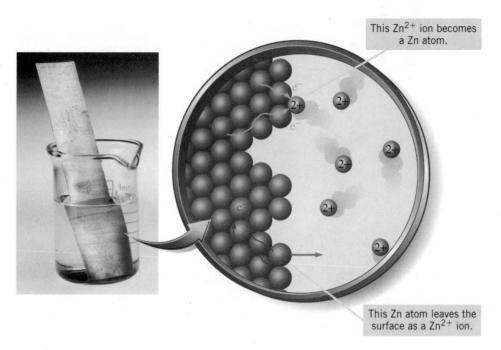

FIGURE 18.3 Electrode equilibrium

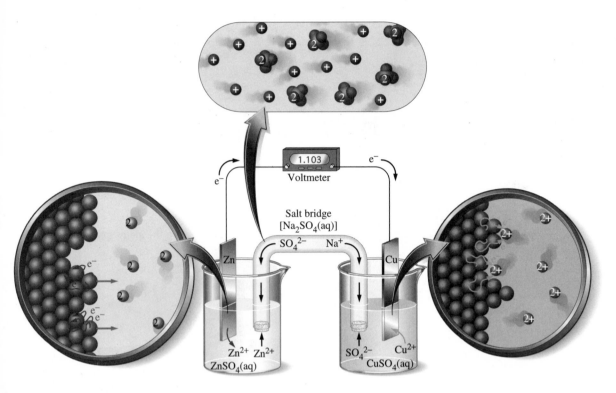

FIGURE 18.4   A zinc-copper voltaic cell

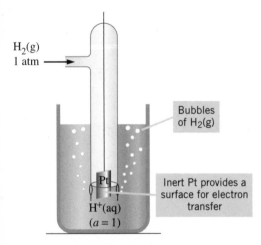

FIGURE 18.6   The standard hydrogen electrode (SHE)

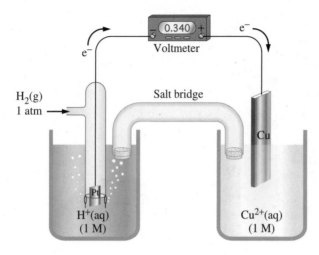

FIGURE 18.7   Measuring the standard potential of the Cu_2+/Cu electrode

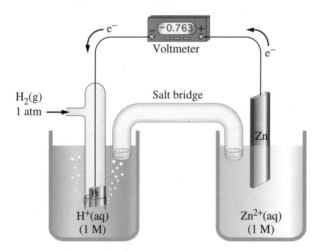

FIGURE 18.8   Measuring the standard potential of the Zn2+/Zn electrode

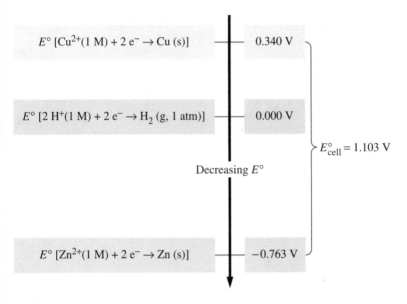

FIGURE 18.9   A representation of standard electrode potentials

Table 18.1 Selected Standard Electrode Potentials at 25 °C			
Reduction Half-Reaction	$E°$, **Volts**	**Reduction Half-Reaction**	$E°$, **Volts**
Acidic Solution		**Acidic Solution**	
$F_2(g) + 2\,e^- \longrightarrow 2\,F^-(aq)$	+2.866	$S(s) + 2\,H^+(aq) + 2\,e^- \longrightarrow H_2S(g)$	+0.14
$O_3(g) + 2\,H^+(aq) + 2\,e^- \longrightarrow O_2(g) + H_2O(l)$	+2.075	$2\,H^+(aq) + 2\,e^- \longrightarrow H_2(g)$	0
$S_2O_8{}^{2-}(aq) + 2\,e^- \longrightarrow 2\,SO_4{}^{2-}(aq)$	+2.01	$Pb^{2+}(aq) + 2\,e^- \longrightarrow Pb(s)$	−0.125
$H_2O_2(aq) + 2\,H^+(aq) + 2\,e^- \longrightarrow 2\,H_2O(l)$	+1.763	$Sn^{2+}(aq) + 2\,e^- \longrightarrow Sn(s)$	−0.137
$MnO_4{}^-(aq) + 8\,H^+(aq) + 5\,e^- \longrightarrow Mn^{2+}(aq) + 4\,H_2O(l)$	+1.51	$Co^{2+}(aq) + 2\,e^- \longrightarrow Co(s)$	−0.277
$PbO_2(s) + 4\,H^+(aq) + 2\,e^- \longrightarrow Pb^{2+}(aq) + 2\,H_2O(l)$	+1.455	$Fe^{2+}(aq) + 2\,e^- \longrightarrow Fe(s)$	−0.440
$Cl_2(g) + 2\,e^- \longrightarrow 2\,Cl^-(aq)$	+1.358	$Zn^{2+}(aq) + 2\,e^- \longrightarrow Zn(s)$	−0.763
$Cr_2O_7{}^{2-}(aq) + 14\,H^+(aq) + 6\,e^- \longrightarrow 2\,Cr^{3+}(aq) + 7\,H_2O(l)$	+1.33	$Al^{3+}(aq) + 3\,e^- \longrightarrow Al(s)$	−1.676
$MnO_2(s) + 4\,H^+(aq) + 2\,e^- \longrightarrow Mn^{2+}(aq) + 2\,H_2O(l)$	+1.23	$Mg^{2+}(aq) + 2\,e^- \longrightarrow Mg(s)$	−2.356
$O_2(g) + 4\,H^+(aq) + 4\,e^- \longrightarrow 2\,H_2O(l)$	+1.229	$Na^+(aq) + e^- \longrightarrow Na(s)$	−2.713
$2\,IO_3{}^-(aq) + 12\,H^+(aq) + 10\,e^- \longrightarrow I_2(s) + 6\,H_2O(l)$	+1.20	$Ca^{2+}(aq) + 2\,e^- \longrightarrow Ca(s)$	−2.84
$Br_2(l) + 2\,e^- \longrightarrow 2\,Br^-(aq)$	+1.065	$K^+(aq) + e^- \longrightarrow K(s)$	−2.924
$NO_3{}^-(aq) + 4\,H^+(aq) + 3\,e^- \longrightarrow NO(g) + 2\,H_2O(l)$	+0.956	$Li^+(aq) + e^- \longrightarrow Li(s)$	−3.040
$Ag^+(aq) + e^- \longrightarrow Ag(s)$	+0.800		
$Fe^{3+}(aq) + e^- \longrightarrow Fe^{2+}(aq)$	+0.771	**Basic Solution**	
$O_2(g) + 2\,H^+(aq) + 2\,e^- \longrightarrow H_2O_2(aq)$	+0.695	$O_3(g) + H_2O(l) + 2\,e^- \longrightarrow O_2(g) + 2\,OH^-(aq)$	+1.246
$I_2(s) + 2\,e^- \longrightarrow 2\,I^-(aq)$	+0.535	$OCl^-(aq) + H_2O(l) + 2\,e^- \longrightarrow Cl^-(aq) + 2\,OH^-(aq)$	+0.890
$Cu^{2+}(aq) + 2\,e^- \longrightarrow Cu(s)$	+0.340	$O_2(g) + 2\,H_2O(l) + 4\,e^- \longrightarrow 4\,OH^-(aq)$	+0.401
$SO_4{}^{2-}(aq) + 4\,H^+(aq) + 2\,e^- \longrightarrow 2\,H_2O(l) + SO_2(g)$	+0.17	$2\,H_2O(l) + 2\,e^- \longrightarrow H_2(g) + 2\,OH^-(aq)$	−0.828
$Sn^{4+}(aq) + 2\,e^- \longrightarrow Sn^{2+}(aq)$	+0.154		

TABLE 18.1   Selected standard electrode potentials at 25 °C

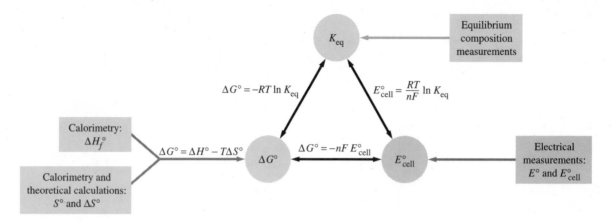

FIGURE 18.12 Summarizing important relationships from thermodynamic, equilibrium, and electrochemistry

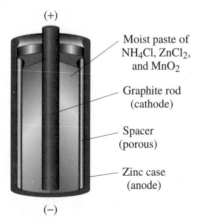

(+)

Moist paste of
NH_4Cl, $ZnCl_2$,
and MnO_2

Graphite rod
(cathode)

Spacer
(porous)

Zinc case
(anode)

(−)

FIGURE 18.16 Cross section of a Leclanche (dry) cell

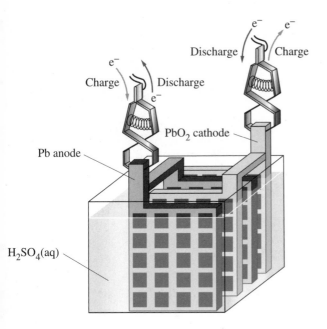

FIGURE 18.17   A lead-acid (storage) cell

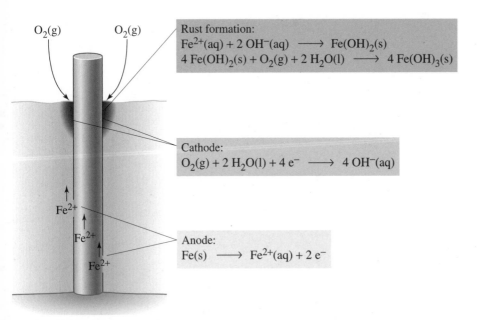

FIGURE 18.19   Corrosion of an iron piling: An electrochemical process

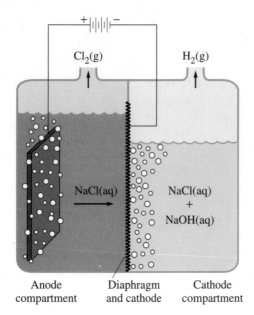

Anode
compartment

Diaphragm
and cathode

Cathode
compartment

FIGURE 18.22 A diaphragm chlor-alkali cell

Chapter 19

Nuclear Chemistry

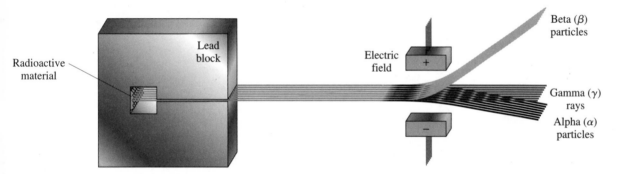

FIGURE 19.1 Three types of radiation from radioactive materials

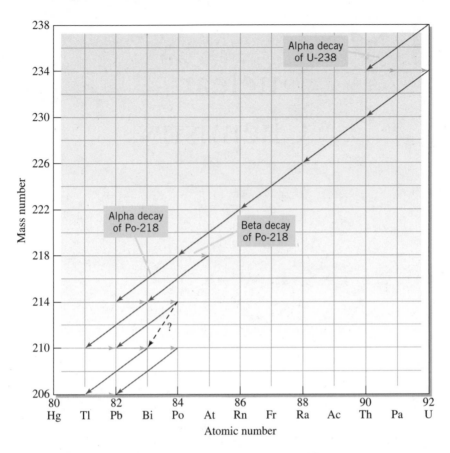

FIGURE 19.2 The natural radioactive decay series for U_{92}^{238}

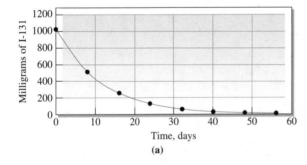

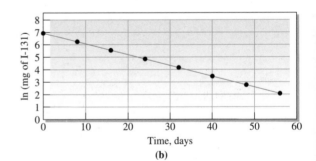

FIGURE 19.3 Graphic representation of decay of I-131

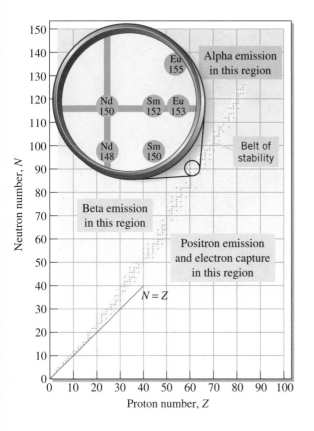

FIGURE 19.5   Neutron-proton ratio and the stability of nuclides

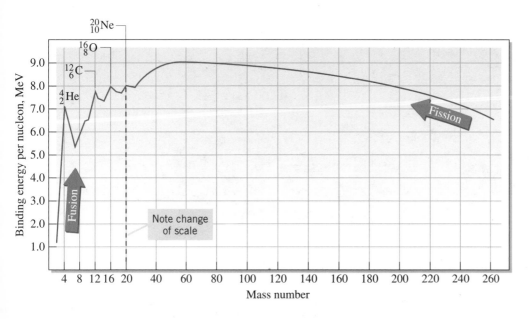

FIGURE 19.7   Average binding energy per nucleon as a function of mass number

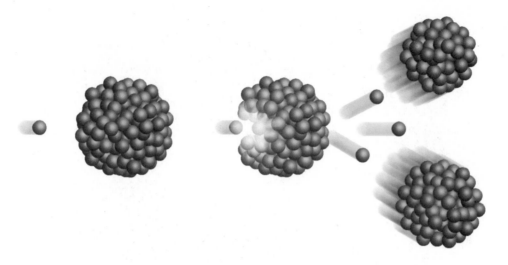

FIGURE 19.8 Nuclear fission of a uranium-235 nucleus with thermal neutrons

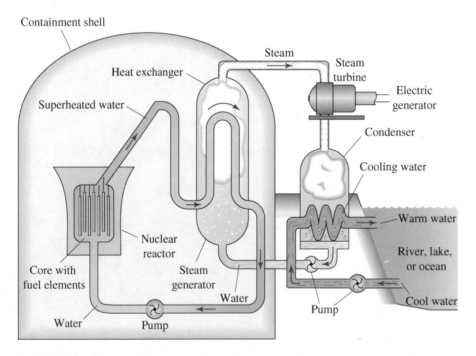

FIGURE 19.9 Schematic diagram of a nuclear power plant used to generate electricity

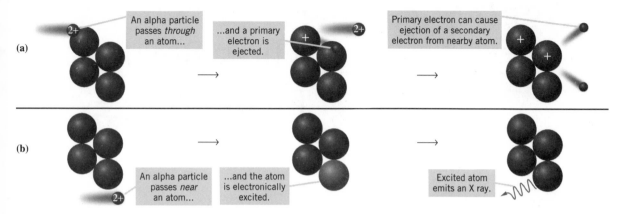

FIGURE 19.11   Some effects of ionizing radiation

Chapter 20

The *s*-Block Elements

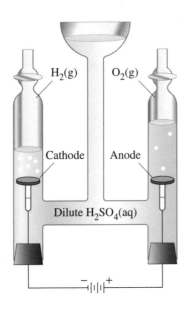

FIGURE 20.2 The electrolysis of water

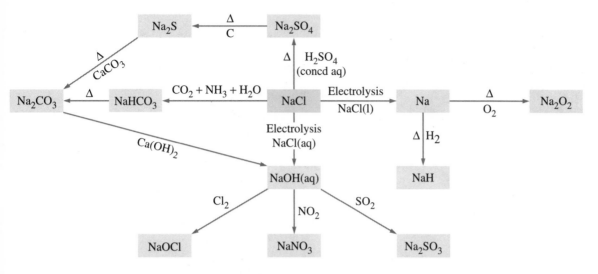

FIGURE 20.5   Preparation of sodium compounds from NaCl

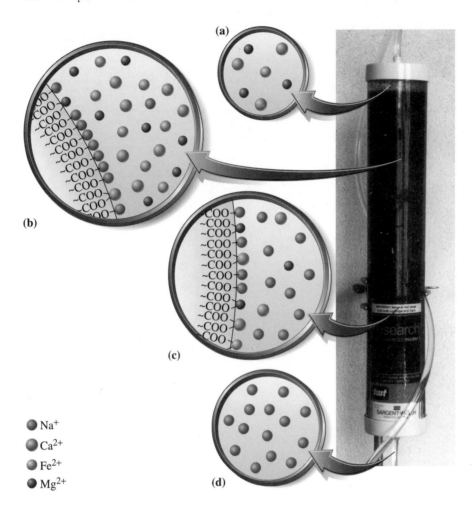

FIGURE 20.9 Water softening by ion exchange

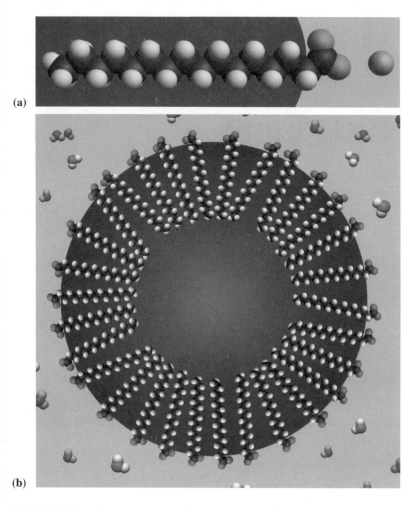

(a)

(b)

FIGURE 20.11   Cleaning action of a soap visualized

Chapter 21

The *p*-Block Elements

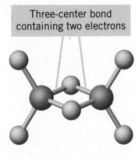

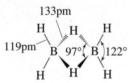

FIGURE 21.1   Structure of diborane, B_2H_6

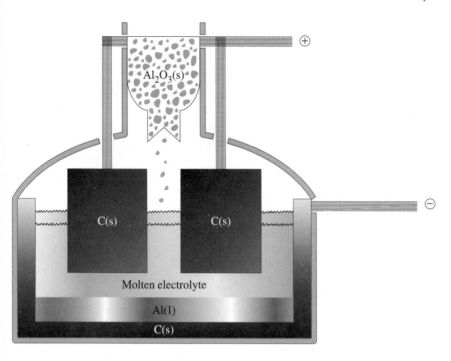

FIGURE 21.2   Electrolysis cell for aluminum production

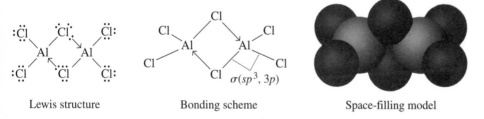

Lewis structure   Bonding scheme   Space-filling model

FIGURE 21.3   Bonding in Al$_2$Cl$_6$

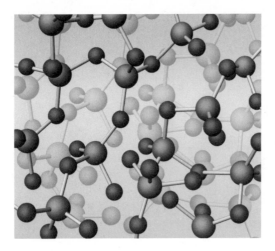

FIGURE 21.5   The structure of silica, SiO_2

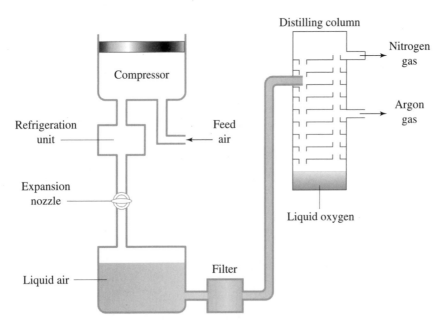

FIGURE 21.8   The fractional distillation of liquid air

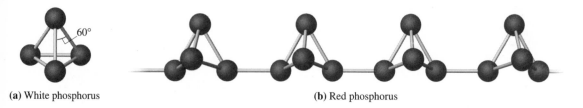

(a) White phosphorus   **(b)** Red phosphorus

FIGURE 21.9   The molecular structures of white and red phosphorus

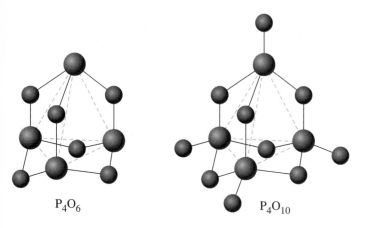

P_4O_6 P_4O_{10}

FIGURE 21.10   Molecular structure of P_4O_6 and P_4O_{10}

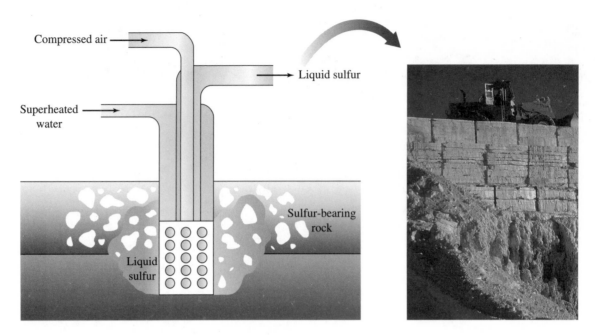

FIGURE 21.13 The Frasch process for mining sulfur

Chapter 22

The *d*-Block Elements and Coordination Chemistry

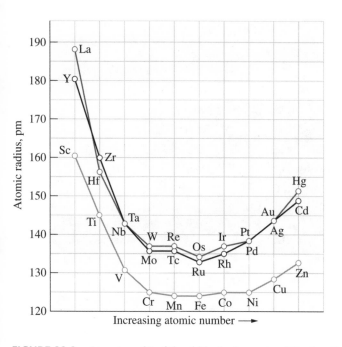

FIGURE 22.2 Atomic radii of the *d*-block elements of the fourth, fifth, and sixth periods

(a) CrO_4^{2-}

(b) $Cr_2O_7^{2-}$

FIGURE 22.4 Chromate and dichromate ions

$Cr[(H_2O)_6]^{3+}$

$[Cr(OH)_4(H_2O)_2]^-$

FIGURE 22.5 Amphoterism of $Cr(OH)_3(s)$

CoCl$_3 \cdot$ 6NH$_3$ CoCl$_3 \cdot$ 5NH$_3$

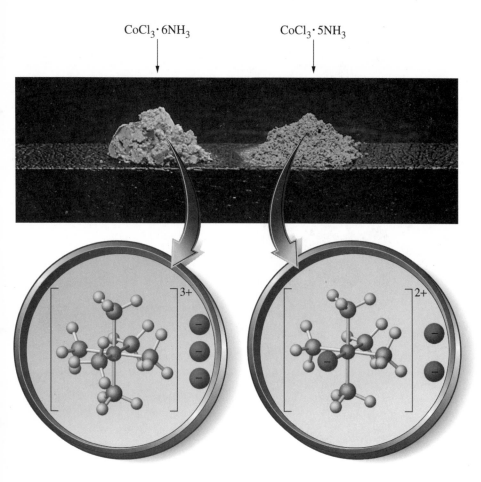

FIGURE 22.10   Two coordination compounds of cobalt(III)

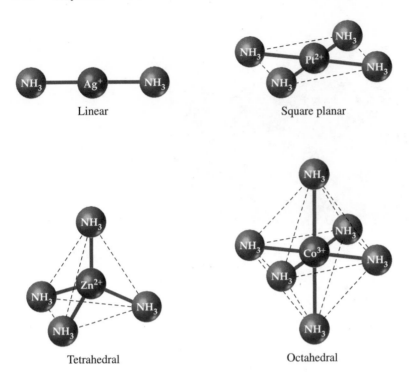

FIGURE 22.11 Four common structures of complex ions

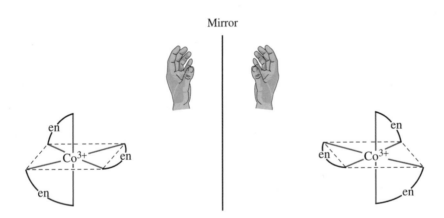

FIGURE 22.15 Optical isomers

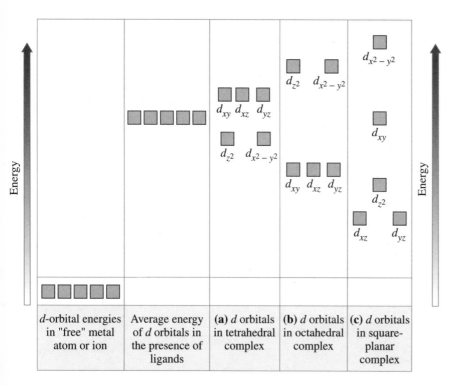

FIGURE 22.18   Schematic representation of *d*-level splitting

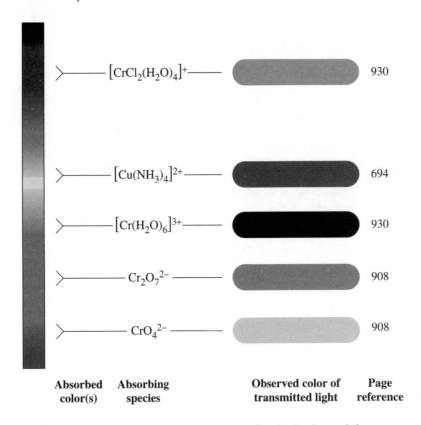

Absorbed color(s)	Absorbing species	Observed color of transmitted light	Page reference
	$[CrCl_2(H_2O)_4]^+$		930
	$[Cu(NH_3)_4]^{2+}$		694
	$[Cr(H_2O)_6]^{3+}$		930
	$Cr_2O_7^{2-}$		908
	CrO_4^{2-}		908

FIGURE 22.21   The relationship between an absorbed color and the corresponding observed color of the transmitted light

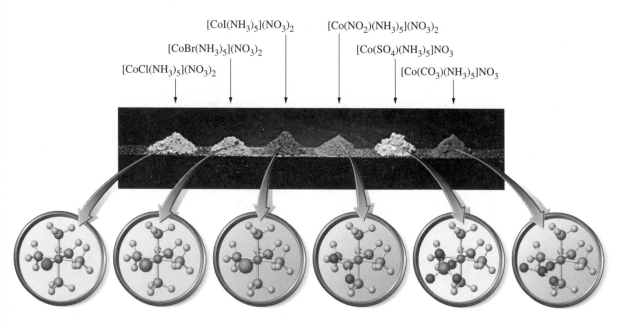

FIGURE 22.22   How various ligands may affect colors of coordination compounds

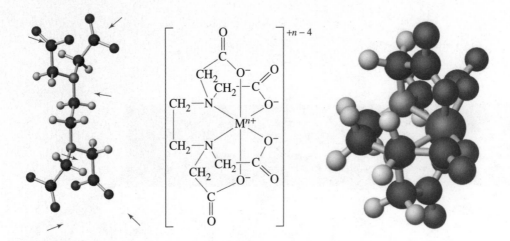

FIGURE 22.24 Structure of a metal-EDTA complex

Chapter 23

Chemistry and Life: More on Organic, Biological, and Medicinal Chemistry

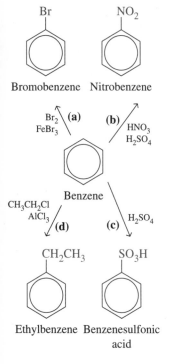

FIGURE 23.1 Some electrophilic aromatic substitution reactions of benzene (a) Halogenation, (b) nitration, (c) sulfonation, and (d) alkylation

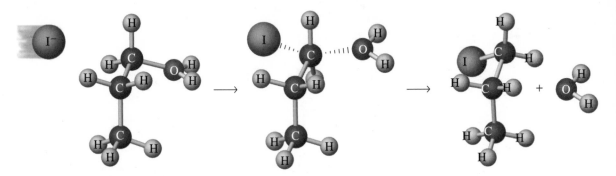

FIGURE 23.2   An SN$_2$ Reaction

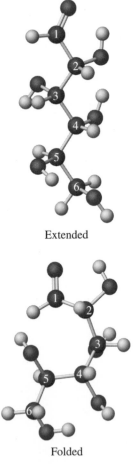

Extended

Folded

FIGURE 23.4   Two arrangements of the open-chain glucose molecule

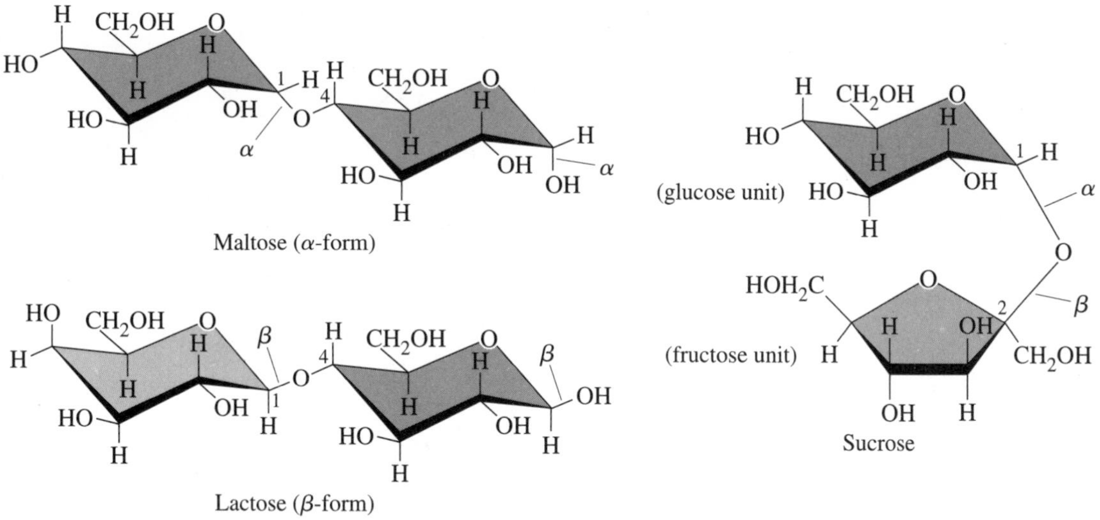

Maltose (α-form)

(glucose unit)

(fructose unit)

Lactose (β-form)

Sucrose

FIGURE 23.5   Three disaccharides

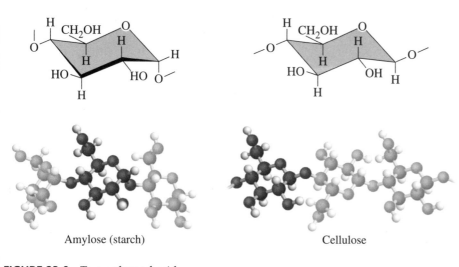

Amylose (starch)

Cellulose

FIGURE 23.6   Two polysaccharides

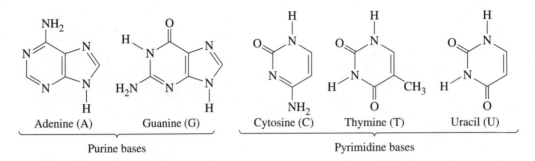

Adenine (A)   Guanine (G)   Cytosine (C)   Thymine (T)   Uracil (U)

Purine bases   Pyrimidine bases

FIGURE 23.9 The five heterocyclic bases found in nucleotides

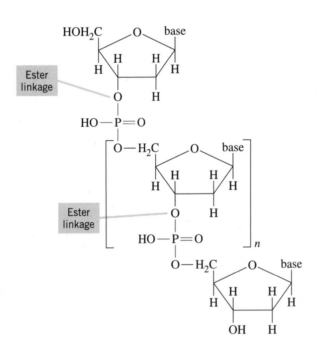

FIGURE 23.10 The backbone of a deoxyribonucleic acid molecule

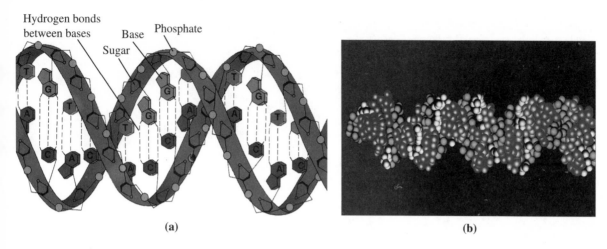

FIGURE 23.11   The DNA double helix

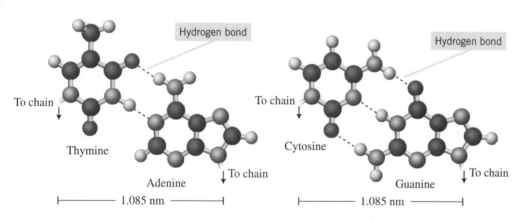

FIGURE 23.12   The pairing of bases in the DNA double helix

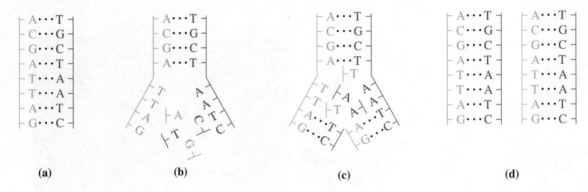

(a)   (b)   (c)   (d)

FIGURE 23.13   DNA

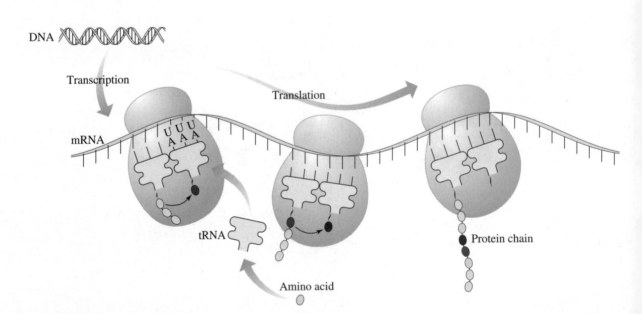

FIGURE 23.15   Protein synthesis visualized: From DNA through mRNA and tRNA to protein chain

FIGURE 23.16   Schematic diagram of the essential components of a simple spectrometer

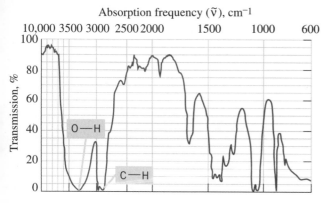

FIGURE 23.18   The infrared spectrum of ethanol, CH_3CH_2OH

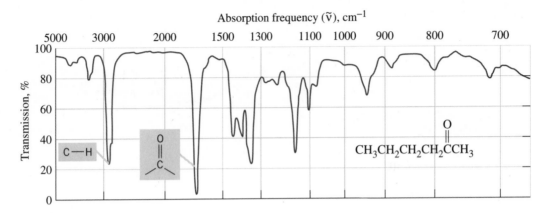

FIGURE 23.19   The infrared spectrum of 2-hexanone

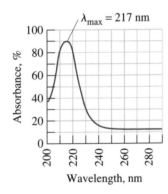

Wavelength, nm   **FIGURE 23.21**   The ultraviolet spectrum of 1,3-butadiene

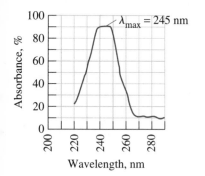

FIGURE 23.22   The ultraviolet absorption spectrum of 1,3-butadiene

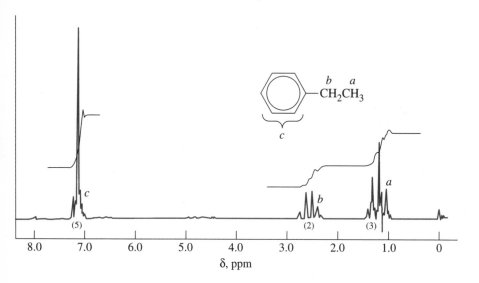

FIGURE 23.25   The NMR spectrum of ethylbenzene

Chapter 24

Chemistry of the Materials: Bronze Age to Space Age

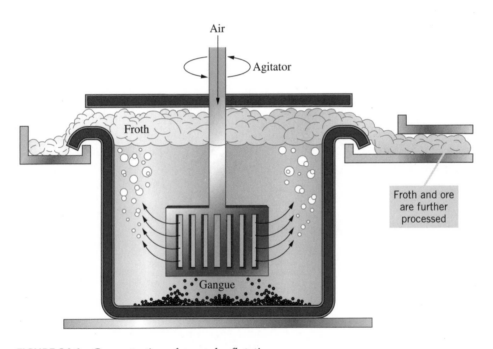

FIGURE 24.1   Concentration of an ore by flotation

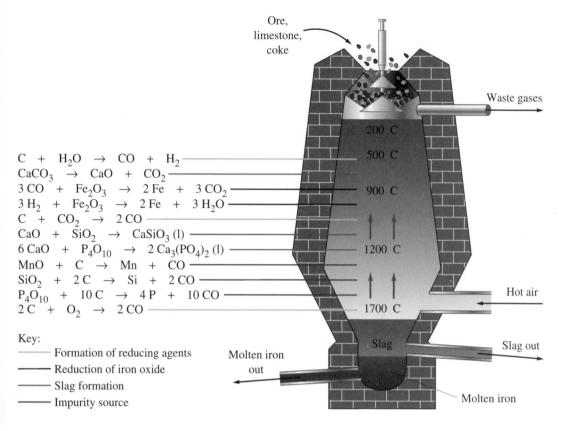

$C + H_2O \rightarrow CO + H_2$
$CaCO_3 \rightarrow CaO + CO_2$
$3\,CO + Fe_2O_3 \rightarrow 2\,Fe + 3\,CO_2$
$3\,H_2 + Fe_2O_3 \rightarrow 2\,Fe + 3\,H_2O$
$C + CO_2 \rightarrow 2\,CO$
$CaO + SiO_2 \rightarrow CaSiO_3\ (l)$
$6\,CaO + P_4O_{10} \rightarrow 2\,Ca_3(PO_4)_2\ (l)$
$MnO + C \rightarrow Mn + CO$
$SiO_2 + 2\,C \rightarrow Si + 2\,CO$
$P_4O_{10} + 10\,C \rightarrow 4\,P + 10\,CO$
$2\,C + O_2 \rightarrow 2\,CO$

Key:
——— Formation of reducing agents
——— Reduction of iron oxide
——— Slag formation
——— Impurity source

FIGURE 24.2   A modern blast furnace

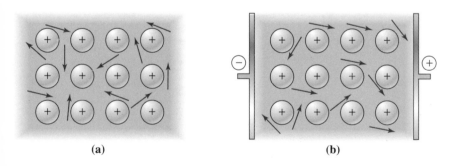

(a)   (b)

FIGURE 24.4   The free-electron model

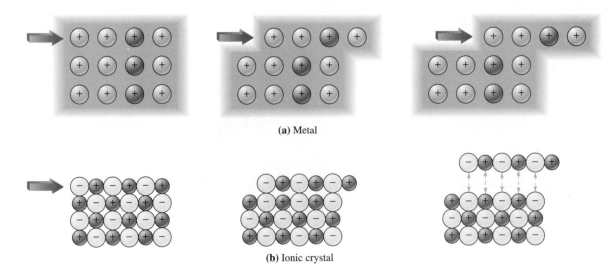

(a) Metal

(b) Ionic crystal

FIGURE 24.5 Deformation of a metal compared to an ionic solid

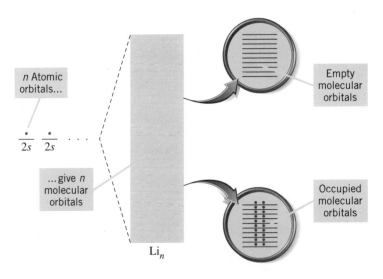

FIGURE 24.7 The 2s band in lithium metal

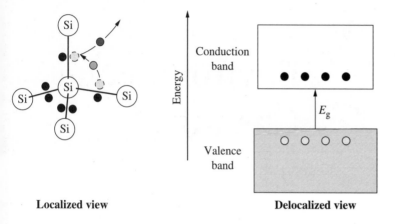

Localized view   Delocalized view

FIGURE 24.11   Two views describing electrical conductivity in semiconductors

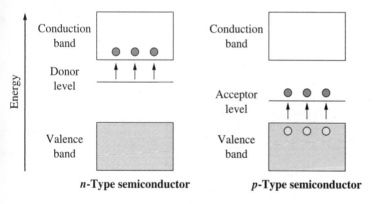

n-Type semiconductor   *p*-Type semiconductor

FIGURE 24.12   Bands in *p*- and *n*-type semiconductors

Chapter 25
Environmental Chemistry

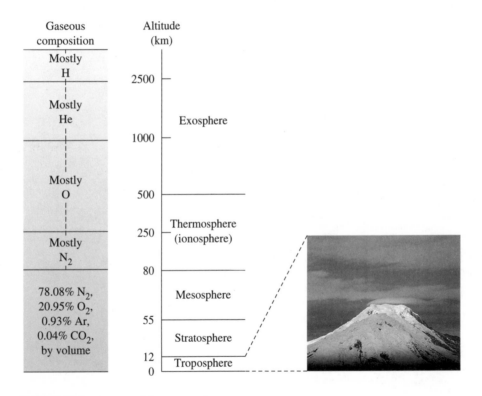

FIGURE 25.1 Layers of the atmosphere

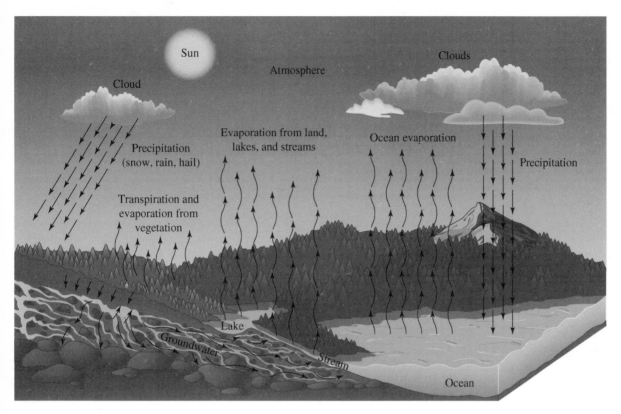

FIGURE 25.2   The hydrologic (water) cycle

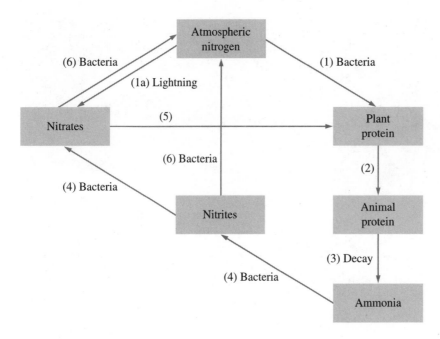

FIGURE 25.5   The nitrogen cycle

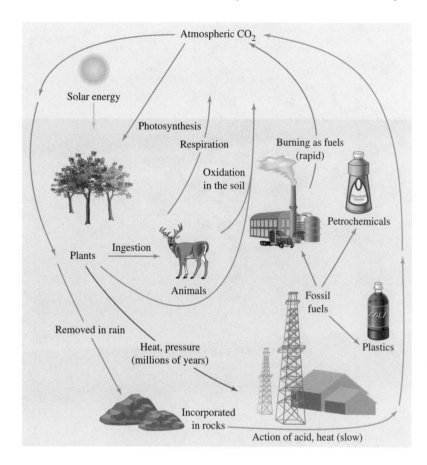

Main cycle
Fossilization tributary
Disruption by human
activities

FIGURE 25.6   The carbon cycle

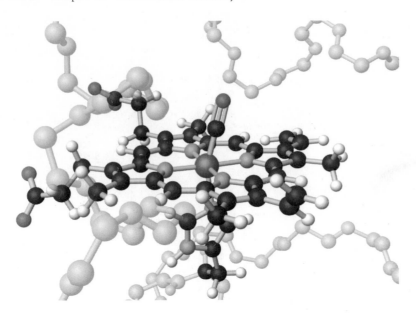

FIGURE 25.7   A molecular view of carbon monoxide poisoning

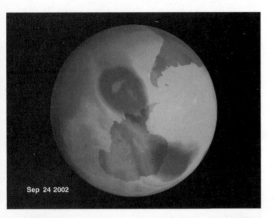

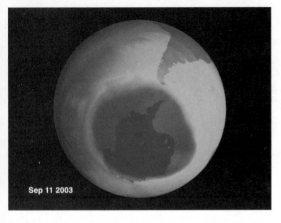

FIGURE 25.11   The ozone hole

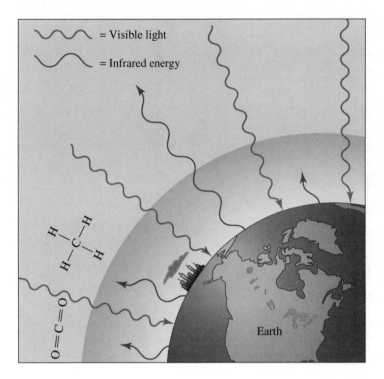

FIGURE 25.12   The greenhouse effect

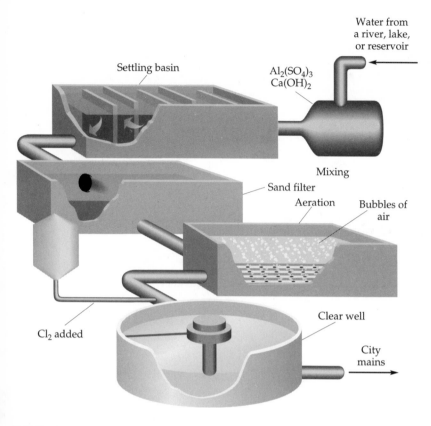

FIGURE 25.14   A diagram of a municipal water purification plant

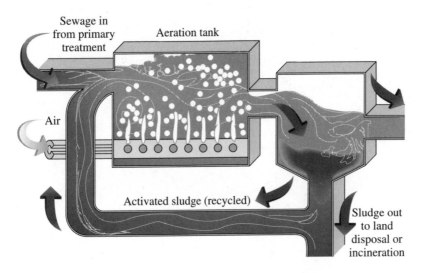

FIGURE 25.15 A diagram of a secondary sewage treatment plant that uses the activated sludge method

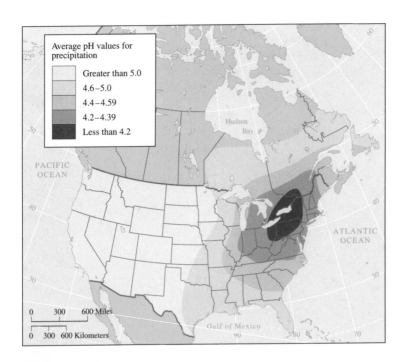

FIGURE 25.16 Acid rain